S0-AKC-482

COMPUTER-AIDED CIRCUIT ANALYSIS USING SPICE

WALTER BANZHAF

Ward College of Technology
University of Hartford
West Hartford, Connecticut

Prentice Hall, Englewood Cliffs, New Jersey 07632

Library of Congress Cataloging-in-Publication Data

Banzhaf, Walter.
    Computer-aided circuit analysis using SPICE / Walter Banzhaf.
        p.   cm.
    Bibliography: p.
    Includes index.
    ISBN 0-13-162579-9
    1. Electric circuit analysis—Data processing.   I. Title.
TK454.B33 1989
621.3815'3'0285—dc19   88-22459
                                                              CIP

Editorial/production supervision: *Mary Carnis*
Cover design: *Wanda Lubelska*
Manufacturing buyer: *Robert Anderson*
Page layout: *Karen Noferi*

IBM® is a registered trademark of International Business
Machines Corporation. PSpice® is a registered trademark of
MicroSim Corporation. VAX® and DEC® are trademarks
of Digital Equipment Corporation.

 © 1989 by Prentice-Hall, Inc.
A Division of Simon & Schuster
Englewood Cliffs, New Jersey 07632

All rights reserved. No part of this book may be
reproduced, in any form or by any means,
without permission in writing from the publisher.

Printed in the United States of America

10 9 8 7 6 5 4 3 2

ISBN 0-13-162579-9

Prentice-Hall International (UK) Limited, *London*
Prentice-Hall of Australia Pty. Limited, *Sydney*
Prentice-Hall Canada Inc., *Toronto*
Prentice-Hall Hispanoamericana, S.A., *Mexico*
Prentice-Hall of India Private Limited, *New Delhi*
Prentice-Hall of Japan, Inc., *Tokyo*
Simon & Schuster Asia Pte. Ltd., *Singapore*
Editora Prentice-Hall do Brasil, Ltda., *Rio de Janeiro*

CONTENTS

PREFACE

This book was written with the express purpose of making it possible (and easy) for anyone in the electronics field to start using SPICE to analyze electric or electronic circuits in a very short time. SPICE is simple to use when you have learned its rules and syntax. Learning those things was challenging for me and for others I spoke with, since a text on the topic didn't exist. My engineering technology students have been using SPICE on a mainframe computer for years; they were able to master its use by following examples which they received in class. Each example contained a schematic diagram of the circuit to be analyzed, the input file which described the circuit to SPICE, and the output file created by SPICE with a discussion of its contents. After making a few of the classical errors (e.g. typing a letter "O" when the digit "0" (zero) was intended), most students were quite comfortable doing analyses completely on their own.

During what was to be a leisurely one-semester sabbatical leave, I had planned to write a small manual on SPICE for use by students at Ward College of Technology of the University of Hartford. Discussions with Greg Burnell, Editor-in-Chief, Electronic Technology, at Prentice-Hall convinced me that there was a need for a text on circuit analysis using SPICE that would appeal to two types of readers.

The first type is engineering and engineering technology students, for whom this book could be a supplemental text for any circuits or electronics course. In addition to learning classical methods of circuit analysis and design, they also need to learn how to perform circuit analysis on computers in order to be prepared for industry. One added benefit of simulating circuit behavior on a computer is that students can gain great insight into circuit behavior which would be otherwise unattainable due to the sheer tedium of the mathematical operations necessary. For example, a student having spent the better part of an hour analyzing a transistor amplifier with a BJT whose beta is 80 is not likely to repeat the exercise with the beta changed to 40 just to satisfy intellectual curiosity. With SPICE, repeating the analysis would take only a few seconds. Also, by using computer simulations to check the work they do by hand students gain confidence in their own analytical abilities.

The second type includes practicing engineers, technologists and technicians who are already competent at circuit analysis and design and want to learn how to use SPICE. This book will help them to use the computer to confirm results obtained by hand, to save time in the laboratory and to go from a concept to a working prototype with minimum delay. Although SPICE (or variations of it) can be found in a large number of universities and industries, it is still not easy to learn to use it due to the lack of a text with sufficient clear examples.

Thus, to answer a clear need, the book you are looking at was developed (the sabbatical was not altogether leisurely). It has a large number of examples which illustrate nearly all the capabilities of SPICE. A first year student may be able to use only the DC and AC analysis examples at first. As that student's knowledge of electronics advances, more and more of the book will be appropriate and useful. Conversely, a practicing engineer, technologist or technician may wish to jump to the more advanced topics right away. It is advisable for that reader at least to skim through the preliminary chapters to see first the pitfalls (not many, but potentially fatal) and second the extensive capabilities of SPICE.

Chapter 1 is an introduction to computer-aided circuit analysis, including a brief history and some insight into how it is done by a computer. The history of SPICE is covered in Chapter 2, emphasizing why such a tool was needed for integrated circuit development.

The specific steps needed to run SPICE are presented in Chapter 3, including the pitfalls you may encounter when creating input files. Chapter 3 also shows how to describe resistors, capacitors, self and mutual inductors, and independent voltage and current sources to SPICE. The actual "how to do it" process is shown in Chapter 4, which includes four example analyses which illustrate the capabilities of SPICE. After reading Chapter 4 the reader should be able to begin using SPICE (assuming you have access to a mainframe computer with SPICE on it, or to a personal computer and a PC version of SPICE).

Linear dependent (controlled) sources are explained in Chapter 5. Chapter 6 is an examination of ten kinds of analysis of which SPICE is capable, containing examples of DC, AC, transient, operating point, transfer function, sensitivity, distortion, noise, Fourier and temperature analysis. How to convince SPICE to make graphs and tables of analysis results is covered in Chapter 7.

The large topic of semiconductors and their SPICE models is shared between two chapters. How to describe semiconductors to SPICE is covered in Chapter 8, while an introduction to changing semiconductor models follows in Chapter 11.

Chapter 9 presents the use of subcircuits in SPICE input files. Transmission lines are examined in Chapter 10, including examples showing time-domain reflectometry and steady-state AC analysis.

The part of the book that ties it all together, Chapter 12, presents 30 examples of SPICE analysis. Each example contains a schematic diagram of a circuit to be analyzed, an input file submitted to SPICE for analysis, the results in an output file created by SPICE and a discussion of key points illustrated by the example.

Five appendices present other SPICE features, how to model transformers and switches, nonlinear (polynomial) dependent sources, a bibliography and sources of SPICE-based circuit analysis software.

Each example in the text can be considered to be a practice problem, and it is expected that readers will have no shortage of circuits to analyze from courses being taken (if students) or from daily work (if engineers, technologists or technicians). Only by applying SPICE to your own circuits will you achieve a mastery of SPICE for your purposes.

I am grateful to all those at the University of Hartford who supported this effort and helped make it happen. This certainly includes my students from whom I learn so much about learning. Greg Burnell and Mary Carnis of Prentice Hall have earned my respect and gratitude for their gentle but effective guidance throughout the writing of the text. The comments provided by James Morris of T.J. Watson School of S.U.N.Y., David O'Brien of Wentworth Institute of Technology, and Carl Zimmer of Arizona State University were most helpful. Without the patience and help of my wife, Mattie (an eagle-eyed proofreader), and my children Amy and Jeremy I could not have written this book. Thank you all.

Walter Banzhaf
Ward College of Technology
University of Hartford
West Hartford, Connecticut

Chapter 1

INTRODUCTION TO COMPUTER-AIDED CIRCUIT ANALYSIS

1.1 THE NEED FOR COMPUTER-AIDED CIRCUIT ANALYSIS

Circuit analysis is a necessary part of circuit design. Once a design for a circuit has been determined, the soundness of the design must be tested to ensure that the circuit does indeed perform as required. Often this involves testing for DC operating point and performance with signal applied, over a range of DC supply voltages, input signal levels, and temperatures. The time-honored way to do this was to build a prototype of the circuit, send it off to the laboratory, and invest large amounts of time and money in putting the circuit through its paces, in the hope that it would perform as desired.

Even if it did, one did not know how the circuit would perform with active devices whose parameters ranged considerably from a nominal value. For example, a small signal transistor typically has a DC current gain, or beta, around 100. The values of beta one might encounter in a lot of acceptable transistors shipped by a reputable manufacturer could range from 40 to 250. A prudent circuit designer would

1

have to test the circuit operation by building many test circuits, using transistors at both ends of the beta range, and putting all of them through their paces. If there were a way to simulate accurately and with confidence the performance of electronic circuits without having to build them, the development cost and time to bring a new circuit into production could be reduced. This is why a great deal of effort was put into simulation of circuit performance by computers.

1.2   ECAP IS DEVELOPED

Computer-aided circuit analysis first became popular in the mid 1960s when the computer program ECAP (Electric Circuit Analysis Program) was developed and made available. This software for mainframe computers (since that's all there were at the time) made it possible for an engineer to analyze an electronic circuit without having to write equations defining the circuit. ECAP was a major accomplishment by programmers at IBM Corporation. Once the program user described the circuit using nodal notation, the program itself wrote the circuit equations and solved them.

Prior to the development of ECAP, digital computers had been in use for years for solving circuit equations. But the tedium involved in writing those equations for all but the most simple circuits was substantial. And once the equations were written by an engineer and solved by a computer, any change to the original circuit design necessitated the rewriting of the circuit equations.

1.3   AFTER ECAP

ECAP was followed by several similar programs, which offered a variety of improvements to ECAP. Among these were SPECTRE, TRAC, NET and CIRCUS. This book is about SPICE, which is the most popular computer program in the world today for predicting the behavior of electronic circuits.

It is not necessary to understand the algorithms and mathematical techniques that are utilized by SPICE to analyze circuits in order to be successful at solving circuit problems with SPICE, any more than one must have a working knowledge of internal combustion engines to drive a car. However, it is vitally important for anyone using SPICE (or any other computer analysis/design program) to be a competent practitioner in the field. Only in that way can one check the correctness and reasonableness of the answers by estimation using sound engineering judgement and/or by working a sample problem through by conventional means.

Also, for some of the trickier kinds of circuit analysis involving circuits which have positive feedback (oscillators, Schmitt triggers) or when non-linear circuit elements are present, a little insight into how SPICE does things can be useful.

1.4   HOW IT'S DONE

If you do want to know how circuit analysis is performed by a computer, a brief look at some of the techniques used by SPICE may give you some insight into the process. A vastly more detailed reference is L. Nagel's thesis, "SPICE2: A Computer Program to Simulate Semiconductor Circuits." Information on ordering a copy can be found in Appendix E. A condensed explanation of how SPICE works is nicely presented by W. Blume in an article in BYTE magazine, referenced in Appendix D.

Circuits are described to SPICE by use of an input file, which lists each circuit element (resistor, capacitor, inductor, voltage and/or current source, semiconductor device) and indicates how each is connected using node numbers. In addition, there are lines in the input file which designate the frequency of sources, temperature, the types of analyses to be done and how the analysis results are to be presented.

SPICE reads the input file and figures out all the circuit elements connected to each node. It then uses Kirchhoff's current law to create a system of equations for the circuit, where the voltages at each node are the unknowns, and the admittance (inverse of impedance) of each branch connecting two nodes are the known quantities. This is a form of nodal analysis. Of course, before this can be accomplished SPICE has to know the model for each semiconductor in the circuit (it does) and determine the specific values for each parameter in the model.

What then results is a system of simultaneous equations, whose size is determined by the number of nodes and circuit elements in the circuit; this group of equations is made into an admittance matrix. In order to solve this matrix, a technique called the Newton-Raphson method is used.   Nagel's thesis not only describes how all the algorithms of SPICE work but also explains alternate methods which were considered, tried and ultimately rejected.   While it makes for interesting reading, as stated earlier it is not necessary to have a particle of insight into the analysis methods used, especially for the person just learning SPICE.

CHAPTER SUMMARY

Computer-aided circuit analysis saves time and money in the process of circuit design.

ECAP, developed by IBM Corporation, was the first widely-available computer program for analyzing electronic circuits. It wrote and solved the equations for a circuit described to it by nodal notation.   ECAP ran on mainframe computers, since personal computers did not exist.

A large number of programs were developed after ECAP;   the most popular is SPICE.

You can use SPICE to predict circuit behavior without having an understanding of how the algorithms in the program function.   However, it is important to be competent in electronic circuit analysis in order to verify results of SPICE.

Chapter 2

HISTORY OF
AND INTRODUCTION
TO SPICE

2.1   ACKNOWLEDGMENT

SPICE was developed by the CAD (Computer-Aided Design) Group of the University of California, Berkeley, and is the sole property of the Regents of the University of California.   The author is grateful to the University of California for granting permission to use excerpts from its Electronics Research Laboratory memoranda in this book.

2.2   THE NEED FOR CAD
WITH INTEGRATED CIRCUITS

During the early 1970s, the number of active devices on integrated circuits (ICs) being developed grew dramatically; SSI (small-scale integration, characterized by about 20 transistors on each IC) led to MSI (medium-scale integration, with about 100 transistors per IC).   (Specific quantitative definitions of SSI, MSI, LSI and VLSI vary; the transistor counts shown here are representative.)   Later came LSI (large-scale integration) and VLSI (very large-scale integration), in which there were literally thousands of transistors

on one integrated circuit.   A 16-bit microprocessor IC has up-
wards of 50,000 components on one tiny piece of silicon.   Mod-
ern active filter, analog to digital converter and digital
signal processing ICs are similarly complex, and increased
numbers and densities of components on ICs are projected.   In
Chap. 1 the need for and methods of testing ordinary (non-IC)
circuit designs were presented.   Imagine the difficulties
involved in breadboarding an integrated circuit prototype
with hundreds or thousands of transistors, only to discover
during the laboratory test phase that the design is faulty.

CAD is also a necessity for IC development because it is
physically impossible to prototype accurately an integrated
circuit using discrete components.   A collection of individu-
al transistors, diodes, resistors and capacitors connected
together by wires on a circuit design board differs considerably from the same circuit implemented on an integrated circuit for the following reasons:

1. The stray, or parasitic capacitances and inductances
of an IC are much smaller than those of a circuit built
on a circuit design board, due to the tiny dimensions of
conductors and components on the IC.

2. Each discrete component on a circuit made on circuit
design board will have a different temperature, due to
the differing amounts of power dissipated by each resistor, diode and transistor.   On an IC, the close proximity
of components ensures that they are all at essentially
the same temperature.   The temperature differences between IC components, though slight, may be significant in
certain operational amplifier applications.   Since the parameters of electronic devices depend on temperature, the
components on an IC will be much better matched than
those in a circuit made with discrete components.

3. Propagation delay, or the time it takes for a change
in voltage at one point in a circuit to reach another
point, is dependent upon the geometry and physical size
of that circuit.   The high-speed performance of a circuit
made on an IC will bear little resemblance to the performance of that circuit built on a circuit design board
with discrete components.

One might then ask why not prototype a circuit design by con-
structing an IC, testing it, and refining the design based on

test results? The costs of creating ICs in the small quantities such an approach would require are prohibitive. Integrated circuits can only be made inexpensively when the volume is very high. Further, even if one did build a prototype IC, if it did not perform as desired it would be extremely difficult to probe the tiny circuit to discover the cause of its substandard performance.

2.3 SPICE BACKGROUND INFORMATION

SPICE is an acronym for Simulation Program with Integrated Circuit Emphasis. It was developed by the Integrated Circuits Group of the Electronics Research Laboratory and the Department of Electrical Engineering and Computer Sciences at the University of California, Berkeley, California. The person credited with originally developing SPICE is Dr. Lawrence Nagel, whose PhD thesis describes the algorithms and numerical methods used in SPICE. SPICE has undergone many changes since it was first introduced; improvements were and are being made on a continuous basis as bugs are encountered in using it.

It is a large (over 17,000 lines of FORTRAN source code), powerful and extremely versatile industry-standard program for circuit analysis and IC design. A significant number of companies have customized Berkeley SPICE for in-house circuit development work. Many software packages based on SPICE have been developed which include the SPICE2 program from Berkeley as the core for performing circuit analysis. Most of these include useful programs added to make the complete package easier to use. For example, SPICE2 is not interactive, it does not have the capability to reference a library of semiconductor components, and its graphs are made on a line printer using ASCII (American Standard Code for Information Interchange) symbols. Some of the commercially available packages are interactive, do include extensive libraries of parts, and have graphics post-processors which make professional-looking graphs. They range in cost from under $100 to $50,000; some run on a personal computer, while others must run on a large mainframe computer. Information on some of these can be found in Appendix E. SPICE3, written in C language, is also available from Berkeley.

SPICE and other packages based on it will be around and widely used for many years to come. As you will see as you read through this book SPICE, though developed for the design of integrated circuits, can be used to solve a great variety of non-IC circuit problems involving power supplies, three-phase power systems, transmission lines and non-linear components, to name a few.   Anyone performing circuit analysis or design should have a working knowledge of SPICE in order to save time, money, and to gain insight into circuit behavior by answering "what if" questions with a computer simulation.   "What if" questions are frequently not answered if the tedium involved in doing so outweighs the curiosity of the person asking the question.

Students will find SPICE to be an important tool for learning circuit analysis and design, and for testing electronic circuits in ways they could not easily do in most college laboratories.   Faculty must be careful, however, to ensure that students are competent in traditional methods of circuit analysis before exposing them to computer methods.

CHAPTER SUMMARY

SPICE was developed for analyzing IC designs by the CAD Group of UC/Berkeley in the 1970s.

A need for a program like SPICE arose because IC designs cannot be tested by breadboarding discrete components, and the trial-and-error method for ICs is too costly.

SPICE (and derivatives of it) are widely used in industry and universities.   Though developed for IC design, it can be a useful tool both for the student and practicing circuit designer with many non-IC circuits.

SPICE can quickly give answers to "what-if" kinds of questions that would be all but unanswerable without computer-aided circuit analysis.

Chapter 3

GROUND RULES OF SPICE AND BASIC ELEMENT LINE RULES

3.1   INTRODUCTION

In order to perform circuit analysis, one must first have a circuit to analyze. The purpose of this profound statement is to reinforce the idea that the circuit design must be done by a person. SPICE can only simulate how a given circuit will behave; it can't suggest ways to improve the performance of that circuit, and it certainly can't synthesize a circuit when given the design criteria.

SPICE is not an interactive program. The circuit and the types of analysis to be done are described to SPICE in an input file. This input file is created using a text editor on a computer, and is submitted to SPICE for analysis. The results of the analysis (or helpful error messages, if errors existed in the input file) appear in an output file written by SPICE. If the results indicate that the circuit needs to be changed, the circuit designer modifies the circuit, edits the input file, runs SPICE again, and examines the output file. This process can be repeated as needed. Thus, SPICE is seen to be a substitute for the cut-and-try method of testing circuit prototypes in the laboratory until the desired performance is achieved.

In some ways, using SPICE can be like trying to register a car by mail:

Doing Car Registration by Mail	Running SPICE
Fill out a form and write a check.	Create a SPICE input file.
Send the envelope off to the friendly folks at the Motor Vehicle Dept. (MVD).	Run SPICE in batch mode.
Much later, receive large envelope from MVD, contents unknown.	Notice a FORTRAN stop message on monitor, meaning SPICE has created an output file.
Open envelope, read contents.	Examine SPICE output file on monitor.

When you open the MVD envelope you might find a list of 73 reasons why you can't register the car <u>with</u> <u>the</u> <u>form</u> <u>the</u> <u>way</u> <u>it</u> <u>is</u>. The SPICE output file may contain error messages which tell you what is wrong with your input file, or perhaps the analysis was successful but the results are presented differently from the way you'd like to see them.

Just possibly the envelope contains your new license plates, and just possibly the circuit analysis went precisely the way you intended. In fact, you don't know whether you have been successful at either venture until you examine the results. Just as the MVD form can be changed and mailed off again, the input file can be edited and resubmitted to SPICE. Eventually the license plates will arrive and the output file will have all you want in it.

3.2   PROCESS FOR USING SPICE

The process for using SPICE to analyze an electronic circuit is detailed below. Fig. 3.1 is a flowchart showing the 4 necessary steps.

1. Start with a schematic diagram of the circuit, and notate it for SPICE. Notation consists of three steps:

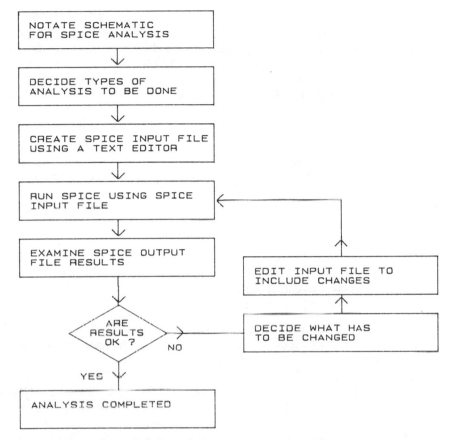

**Figure 3.1 Flowchart Showing How to
Analyze a Circuit with SPICE.**

a. Give each component, or circuit element, a name. For example, a 10K ohm resistor could be named R, R7, RLOAD or RBASEQ1. Notice that the name of a resistor must start with the letter R, and can contain from 1 to 8 characters. Each element must have a different name. Capacitors start with C, inductors begin with L, etc.

b. Assign one node (a point of connection between two or more circuit elements) the node number zero. This is the ground (or datum) node, and all other circuit node voltages will be expressed with respect to this node. While the datum node need not be the actual circuit ground node, it is sensible to assign it this way.

c. Each additional node in the circuit is given a node number, which must be a positive integer. The order in which nodes are numbered is arbitrary, and node numbers need not be sequential. Each node must be connected to at least two elements, except for transmission lines and MOSFET substrates.

2. Decide what type(s) of analysis you want to perform on the circuit. SPICE can do DC, AC, transient, DC transfer function, DC small-signal sensitivity, distortion, noise, and Fourier analyses. Based on the types of analyses to be done, one or more control lines will have to be added to the input file. A control line could specify the values of DC voltage or current for a source, the range of frequencies an AC source is to have, or the time interval over which a transient analysis is to take place and the time step size to be used. Control lines also instruct SPICE how to present the results of the analyses; tabular form and graphical form are available.

3. Create an input file for SPICE using a text editor. Use only capital letters in this file; though some SPICE derivatives will accept lower-case letters, others will not. The first line of the input file MUST be a title line. The last line of the input file MUST be .END . Note that in the .END line the period (.) is a necessary part of the line. Other than the title line and the .END line, the element lines and the control lines can be in any order (one exception to this is when subcircuits are used). Generic and annotated sample SPICE input files are shown in Figs. 3.2 and 3.3, respectively. Additional sample SPICE input files can be found in Chaps. 4, 7, 9, 10 and 12 as well as in Appendices B and C.

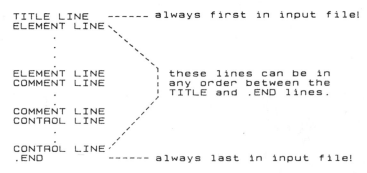

Figure 3.2 Generic SPICE Input File.

ANNOTATED INPUT FILE					*This title line is a must!*
VBATTERY	**11**	**0**	**6**		*A DC source, 6 V, node 11 is +.*
* BATTERY IS A NI-CAD TYPE					*A comment line reminds you later.*
RA		**11**	**6**	**20**	*Element line for 20 ohm resistor.*
RC		**6**	**0**	**51**	*Here's a 51 ohm resistor.*
RB		**6**	**18**	**47**	*47 ohm R between nodes 6 and 18.*
* REMEMBER R TOLERANCE!					*Comments make file readable.*
LOUTPUT		**18**	**0**	**3**	*Element line for 3 H inductor.*
.OP					*Control line; does operating point.*
.END					*This line must be the last one.*

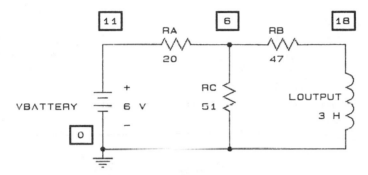

Figure 3.3 (a) Annotated Sample SPICE Input File; (b) Circuit Described by Sample SPICE Input File.

4. Run SPICE, which will perform a circuit analysis and create an output file. Examine the output file. If the results are not satisfactory, revise the circuit as necessary, edit the input file to reflect the change, and run SPICE again.

3.3   RIGID RULES, HANDY HINTS AND USEFUL THINGS TO REMEMBER

1. Each node must have a DC path to ground, which is node zero. This is necessary so that SPICE can do an initial DC small-signal analysis to determine the DC voltage at each node. The simple case where two capacitors are in series, with a node at their junction, will lead to an error message and the aborting of the SPICE analysis. A way to avoid this is to place a large value of resistor (1 giga-ohm or 1 tera-ohm) in parallel with a capacitor, thus providing a DC path to ground. Such a large value

of resistance will not materially alter the circuit performance, and will allow SPICE to run the analysis.

2. SPICE will not allow a loop of voltage sources and/or inductors, nor a series connection of current sources and/or capacitors. The use of a very small resistor in series with an inductor, or a very large resistor in parallel with a capacitor (see hint 1 above) will eliminate this problem.

3. Comment lines are very useful in programs that are to be looked at by a human being, even when that human is the author of the programs. The same is true of comments in SPICE input files. Well-crafted comments will help when one refers back to an old input file to assist in creating a new file.
  To make a comment line, put an asterisk (*) in the first column and write the comment. Comments may appear anywhere in the input file except for the first line (which is always the title line) and the last line (which is always .END). SPICE ignores all comment lines; however, you will be glad you took the time to put them in.

4. Creating the .END line may get you in trouble. Some text editors (EDT on a VAX computer is one) will put in a blank, empty line after the last carriage return (or ENTER). If you type .END and then hit carriage return, an empty, blank line (but nonetheless a line) is put after the .END line. When SPICE examines the input file, it will find that the last line is not .END, and will give you an error message. If you get this disconcerting message, carefully examine the input file to make sure that the very last line is .END and not a blank.

5. Some text editors put "funny" characters in what to you looks like a plain text file. When SPICE examines the input file and encounters these "funny" characters, errors can result. Be sure the text editor you use to create input files is in non-document mode and does not add undesired characters.

6. SPICE will recognize numbers expressed in a variety of ways. Numbers can be integers (-12, 374), floating point (2.123, -54.997), in scientific notation (8.623E2, -3E04,

10.775E-5, where E denotes the power of 10), or can use the following engineering-notation suffixes:

F = 1E-15 (femto)	K = 1E3 (Kilo)
P = 1E-12 (pico)	MEG = 1E6 (Mega)
U = 1E-6 (micro)	G = 1E9 (Giga)
M = 1E-3 (milli)	T = 1E12 (Tera)
MIL = 25.4E-6	

So, 270 may be expressed as 270, 270.0, 2.7E2, or 0.27K. Also, 0.000543 could be represented by 5.43E-4, 0.543M, and 543U. Any letters following one of the suffixes above will be ignored, as will letters that are not suffixes following a number. Thus, 270V, 2.70E2VOLTS, 0.27KOHM, and .27KW all mean the same number, 270; the letters which follow valid numbers (VOLTS, OHM, W) are without meaning to SPICE.

At this point, you may want to glance quickly at the rest of this chapter and then jump ahead to Chap. 4 to see some representative circuits, their SPICE input files, and the resulting SPICE output files. The remainder of this chapter is a detailed presentation of element lines, to which you can return as needed to understand subsequent examples or when writing your own input files.

3.4   RULES FOR ELEMENT LINES

In the sample element lines which follow, XXXXXX, YYYYYY, and ZZZZZZ mean that any alphanumeric string totaling up to seven characters can be used after the letter (R, C, L, etc.) which specifies the type of element. Examples of valid names for a capacitor are C, CFILTER, C472, and CINPUT34. Any data contained within a less-than sign (<) and a greater-than sign (>) represents optional information. Any punctuation, such as equal signs, parentheses and commas, must be included as shown. Extra spaces in an element line are ignored. An element line may be continued by entering a + (plus) sign in the first column of the next line. SPICE will start reading the continued element line in the second column.

SPICE recognizes an element by the first letter of the element name on the element line. There are 15 letters for element names as follows:

C   Capacitor
D   Diode
E   Voltage-Controlled Voltage Source
F   Current-Controlled Current Source
G   Voltage-Controlled Current Source
H   Current-Controlled Voltage Source
I   Independent Current Source
J   Junction Field-Effect Transistor (JFET)
K   Coefficient of Coupling, for Mutual Inductance
L   Inductor
M   Metal-Oxide Semiconductor Field-Effect Transistor (MOSFET)
Q   Bipolar Junction Transistor (BJT)
T   Transmission Line
V   Independent Voltage Source

In this chapter, resistor, capacitor, inductor, mutual inductor and independent current and voltage sources will be presented. These basic circuit elements will allow you to do analyses on a vast array of circuit types. Controlled (or dependent) current and voltage sources are discussed in Chap. 5; semiconductors (diode, BJT, JFET and MOSFET) will be covered in Chap. 8; transmission lines are presented in Chap. 10.

3.4.1 Resistors

RXXXXXX N1 N2 VALUE <TC=TC1<,TC2>>

TC1 and TC2 are optional temperature coefficients; if not specified, they are both assumed to be zero. The value of the resistor as a function of temperature is given by:

$$VALUE(temp) = VALUE(tnom)*(1+TC1(temp-tnom) +$$
$$TC2*(temp-tnom)^2)$$

EXAMPLES:

RINPUT 8 17 3.5E2 TC = -0.004, 0

This describes a resistor between nodes 8 and 17, with a negative and linear temperature coefficient. At the nominal temperature, its resistance is 350 ohms. If the nominal temperature is 27 C, at 127 C the resistance would be given by:

VALUE = 350 * (1 + (-0.004)(127 - 27))
VALUE = 350 * (0.6) = 210 ohms

R12 5 9 2700K
A 2.7 megohm resistor between nodes 5 and 9.

RWIRE 6 31 0.008
An 8 milli-ohm resistor between nodes 6 and 31.

3.4.2 Capacitors

CYYYYYY N+ N- VALUE <IC=INCOND>

N+ and N- are the nodes to which the capacitor is connected; N+ is the positive node and N- is the negative node. IC is the value of voltage across the capacitor at time zero and is optional. If an initial condition is given in the element line for a capacitor, it applies only if the UIC (use initial conditions) option is included on the .TRAN control line. .TRAN causes a transient analysis to be done; this is covered in Chap. 6.4.

EXAMPLES:

CBYPASS 3 0 4.7E-8
This describes a 0.047 microfarad capacitor between nodes 3 and 0.

CFILTER 14 22 33U IC=12V
This describes a 33 microfarad capacitor, connected between nodes 14 and 22, initially charged (at time = 0) to +12 volts DC (node 14 positive compared to node 22).

C3 35 8 9100P
A 9.1 nanofarad capacitor between nodes 35 and 8.

A nonlinear capacitor whose capacitance depends on the instantaneous voltage across it can be described to SPICE by

CXXXXXX N+ N- POLY C0 C1 C2 . . . <IC = INCOND>

where C0, C1, C2 . . . are coefficients of a polynomial which determine the capacitance. The expression for the capacitance (in farads) is

$$C = C0 + C1*V + C2*V^2 + . . .$$

where V is the instantaneous voltage across the capacitor.

3.4.3 Inductors

LZZZZZZ N+ N- VALUE <IC=INCOND>

N+ and N- are the nodes to which the inductor is connected; N+ is the positive node and N- is the negative node. IC is the value of current flowing through the inductor at time zero, and is optional. Positive current is assumed to flow from the N+ node, through the inductor, to the N- node. If an initial condition is given in the element line for an inductor, it applies only if the UIC (use initial conditions) option is included on the .TRAN control line.

EXAMPLES:

LTANK 11 34 5.6E-4
  A 560 microhenry inductor between nodes 11 and 34.

LOSC 2 55 12M
  A 12 millihenry inductor between nodes 2 and 55.

L7 29 41 2.85 IC = 3
  This describes a 2.85 henry inductor between nodes 29 and 41, with a current at time zero of 3 amps flowing from node 29 to node 41 through the inductor.

A nonlinear inductor whose inductance depends on the instantaneous current through it can be described as follows:

LXXXXXX N+ N- POLY L0 L1 L2 . . . <IC = INCOND>

where L0, L1, L2 . . . are coefficients of a polynomial which determine the inductance. The expression for the inductance (in henries) is

$$L = L0 + L1*I + L2*I^2 + . . .$$

where I is the instantaneous current through the inductor.

3.4.4 Mutual Inductors

KXXXXXX LYYYYYY LZZZZZZ VALUE

This is an element line which allows two inductors (each of which is described by its own element line) to be magnetically linked. VALUE is the magnitude of the coefficient of coupling, and must be greater than 0 and must not exceed 1. Notice that this element line has no node numbers, since the two inductors stated in this line each have node connections. The dotted ends of the inductors are assumed by SPICE to be the N+ nodes of each.

EXAMPLES:

KPRISEC LP LS 0.97
    The coefficient of coupling between LP and LS is 0.97.

KAX LANT LXMTR .08
    The coefficient of coupling between LANT and LXMTR is 0.08.

3.5   INDEPENDENT SOURCES

VXXXXXX N+ N-< <<DC> DC/TRAN-VALUE> <AC <ACMAG <ACPHASE>>>
            <TRANKIND(TPAR1 TPAR2 . . .)> >

IXXXXXX N+ N- < <<DC> DC/TRAN-VALUE> <AC <ACMAG <ACPHASE>>>
            <TRANKIND(TPAR1 TPAR2 . . .)> >

The word "independent" in independent source means that the current flowing through the current source, or the voltage across the voltage source, is not dependent on any other circuit parameter. Examples of independent sources are ideal batteries, an electric power outlet (for reasonable load currents), or a function generator with zero output impedance.

The collector-emitter current in a common-emitter transistor circuit can be thought of as a "dependent" source, since the C-E current is controlled by, or is dependent upon, the base current. For a typical small-signal transistor, collector current would be 100 times the base current. This would be a current-controlled current source. A field-effect transistor (FET) would be an example of a voltage-controlled current source, since the drain-source current is controlled by the gate-source voltage.

The letters DC following the N- node number indicate that the source does not vary with time, and are optional. DC/TRAN-VALUE is the magnitude of the source for DC and transient analysis; this value may be left out if it is zero. The letters AC indicate that the source is a sinusoid with magnitude and phase described by ACMAG and ACPHASE when an AC analysis is performed. ACMAG and/or ACPHASE may be omitted; ACMAG will default to one volt, while ACPHASE will default to zero degrees.

In addition, for the purpose of transient analysis only, an independent source can be one of five time-dependent functions (TRANKIND): pulse, sinusoidal, exponential, piece-wise linear, or single-frequency FM. TPAR1 etc. are the parameters associated with the particular TRANKIND. These five time-dependent functions are described in detail later in the chapter.

SPICE considers positive current to be that current which flows from the positive node (N+) to the negative node (N-) through the source. When the product of voltage and current is positive, a source is receiving power from the circuit; if the product is negative, the source is supplying power to the circuit.

While it is possible to have all the options specified for an independent source (DC, AC and TRANKIND), it would be better in most cases for reasons of clarity (and sanity) to use only one option at a time with a source. It is also possible to specify none of the options, in which case the source is dead. A dead voltage source, which acts like a short circuit, has a very useful purpose. See the VBORING example below.

EXAMPLES:

VBATTERY 12 23 DC 6
   This is a DC voltage source, with node 12 positive by
6 V compared to node 23.

ILED 38 46 SIN(20M 80M 10KHZ)
   This describes a current source with a DC value (or
offset) of 20 mA, and a time-dependent value (when a transient analysis is performed) of a 10 KHz sinusoid with a
peak amplitude of 80 mA. Positive current is assumed to
flow from node 38 to node 46, through the current source.

VBORING 7 5
   VBORING is a dead voltage source between nodes 7 and
5; it always has zero volts across it, and behaves like a
short circuit. While it may appear to be useless, it can
be used as an ammeter to measure currents in any circuit
branch desired. SPICE uses dead voltage sources for this
purpose. Note: if the current in VBORING were positive,
that would mean current was flowing from node 7 to node 5
through VBORING.

VFMSIG 29 11 AC 1 45 SFFM(0 2 1K 2.5 200)
   The voltage source VFMSIG can be either of two things,
depending on what type of analysis is being done. When
an AC analysis is done (by a separate control line starting with .AC), VFMSIG is an AC sinusoidal source with a
magnitude of 1 V and a phase angle of 45 degrees. The
frequency is set by the .AC control line. When a transient analysis is performed (by a separate control line
starting with .TRAN), VFMSIG becomes a single-frequency
FM source, with an amplitude of 2 V peak, carrier frequency of 1 KHz, modulation index of 2.5, and modulating (or
signal) frequency of 200 Hz.
   This example shows a two-purpose source that can be described to SPICE with a single element line. In many
cases it would be better to run SPICE twice: first with
the AC source and AC analysis and then with the single-
frequency FM source and transient analysis. To do that,
a text editor would be used to edit the input files so
that one contained VFMSIG 29 11 AC 1 45 and the other
contained VFMSIG 29 11 SFFM(0 2 1K 2.5 200).

VSYNC 8 17 PULSE(-10 20 10M 1U 2U 15M 60M)

VSYNC is a pulse train starting at -10 V, going to 20 V, delayed from time-zero by 10 ms, with a rise time of 1 microsec, fall time of 2 microsec, pulse width of 15 ms, and period of 60 ms.

ITRIANGL 22 34 PWL(0 0 .25 3 .75 -3 1.25 3 1.75 -3 2 0)

This describes two cycles of a triangular current waveform, starting at 0 V and ramping between +3 V and -3 V with a period of 1 s.   Positive current flows from node 22 to node 34 through the current source ITRIANGL.

3.5.1 Time–Dependent Functions for Independent Sources

Five kinds of time-dependent functions are available for transient analysis using independent sources:

1. PULSE - A single pulse or pulse train, with rise and fall times that can be specified. This can be used to produce pulse, squarewave, triangle and sawtooth sources.

2. SIN - A sinusoid in which the start can occur after time zero. It may be exponentially damped if desired.

3. EXP - A single pulse, with exponentially growing rise and exponentially decaying fall; rise and fall may have different time constants.

4. PWL - A piece-wise linear description of a source which can be used to approximate nearly any waveform (if you have the time to generate the time-voltage coordinate list). For example, a complex TV sync waveform or an electrocardiogram can be described to SPICE using PWL.

5. SFFM - Single-frequency FM (frequency modulation) source with independently-specified offset voltage, amplitude, carrier frequency, FM modulation index, and modulation frequency (limited to a single sinusoid).

3.5.1.1 Pulse Function for Independent Sources

```
VXXXXXX    N+ N- PULSE(V1 V2 TD TR TF PW PER)
IXXXXXX    N+ N- PULSE(V1 V2 TD TR TF PW PER)
```

describe a voltage or current pulse source. The parameters are:

Parameter	Default Value	Unit
V1 (initial value)	must specify	Volts or amps
V2 (pulsed value)	must specify	Volts or amps
TD (time delay)	0.0	seconds
TR (rise time)	TSTEP	seconds
TF (fall time)	TSTEP	seconds
PW (pulse width)	TSTOP	seconds
PER (period)	TSTOP	seconds

TSTEP and TSTOP refer to the step size and stop time in the transient analysis which is specified by a .TRAN TSTEP TSTOP control line in the input file. A pulse waveform illustrating the parameters above is shown in Fig. 3.4.

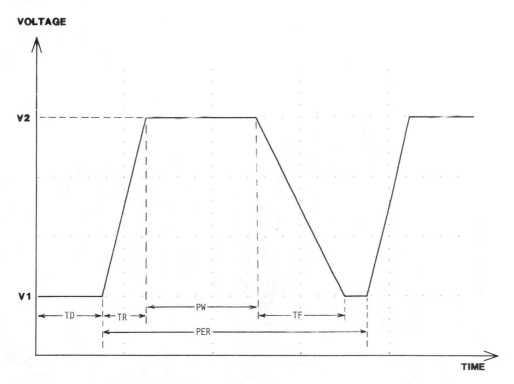

Figure 3.4 SPICE Pulse Waveform.

3.5.1.2 Sinusoidal Function for Independent Sources

```
VXXXXXX    N+ N- SIN(VO VA FREQ TD THETA)
IXXXXXX    N+ N- SIN(VO VA FREQ TD THETA)
```

describe a voltage or current sinusoidal source. The parameters are:

Parameter	Default Value	Unit
VO (offset value)	must specify	Volts or amps
VA (amplitude)	must specify	Volts or amps
FREQ (frequency)	1/TSTOP	Hertz
TD (delay time)	0.0	seconds
THETA(damping factor)	0.0	1/second

TSTOP refers to the stop time in the transient analysis which is specified by a .TRAN TSTEP TSTOP control line in the input file. A damped sinusoidal waveform illustrating the parameters above is shown in Fig. 3.5.

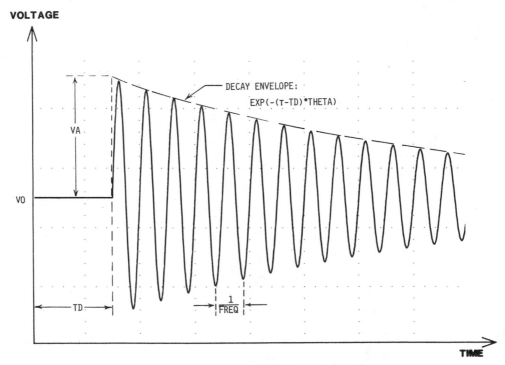

Figure 3.5 SPICE Sinusoidal Waveform.

The waveform as a function of time, t, is described by the relationships:

from t = 0 to TD,   VXXXXXX = VO
from t = TD to TSTOP,   VXXXXXX =

$$VO + VA*exp(-(t-TD)*THETA)*sine(2*PI*FREQ*(t-TD))$$

If TD is not specified, the sine starts at time-zero. If THETA is not specified, the sine has no decay and therefore has constant peak amplitude.

3.5.1.3 Exponential Function for Independent Sources

VXXXXXX   N+ N- EXP(V1 V2 TD1 TAU1 TD2 TAU2)
IXXXXXX   N+ N- EXP(V1 V2 TD1 TAU1 TD2 TAU2)

describe a voltage or current exponential source. The parameters are:

Parameter		Default Value	Unit
V1	(initial value)	must specify	Volts or amps
V2	(pulsed value)	must specify	Volts or amps
TD1	(rise delay time)	0.0	seconds
TAU1	(rise time constant)	TSTEP	seconds
TD2	(fall delay time)	TD1 + TSTEP	seconds
TAU2	(fall time constant)	TSTEP	seconds

TSTEP refers to the step time in the transient analysis which is specified by a .TRAN TSTEP TSTOP control line in the input file. Figure 3.6 shows an exponential waveform illustrating the parameters above.

The waveform as a function of time, t, is described by the relationships:

from t = 0 to TD1,   VXXXXXX = V1
from t = TD1 to TD2,   VXXXXXX =

$$V1 + (V2-V1)*(1-exp(-(t-TD1)/TAU1))$$

from t = TD2 to TSTOP,   VXXXXXX =

$$V1 + (V2-V1)*(1-exp(-(t-TD1)/TAU1)) +$$
$$(V1-V2)*(1-exp(-(t-TD2)/TAU2))$$

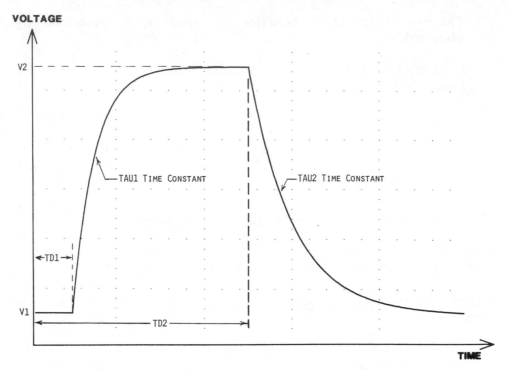

Figure 3.6   SPICE Exponential Waveform.

3.5.1.4 Piece-wise Linear Function for Independent Sources

VXXXXXX   N+ N- PWL(T1 V1 <T2 V2 T3 V3 . . .>)
IXXXXXX   N+ N- PWL(T1 V1 <T2 V2 T3 V3 . . .>)

describe a voltage or current piece-wise linear source. The parameters are:

Parameter	Default Value	Unit
Tn (nth time value)	none	seconds
Vn (nth voltage value)	none	Volts or amps

Any arbitrary voltage or current waveform may be described by a piece-wise linear approximation. Each time and voltage (or current) pair is the coordinate of a point on the graph of the waveform versus time. Time points need not be evenly spaced, and the source value at times between specified points is done by linear interpolation. Figure 3.7 shows a PWL voltage source described by the element lines

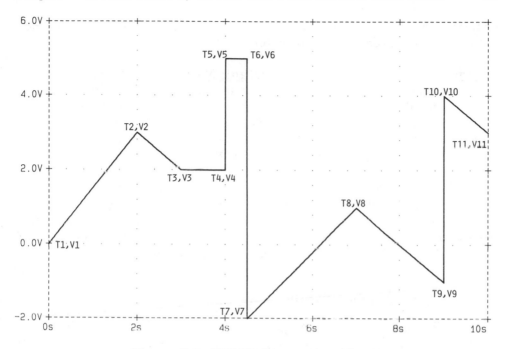

**Figure 3.7   SPICE Piece-wise Linear
Waveform.**

```
VPWL 1 0 PWL(0 0  2 3  3 2  4 2  4.01 5
+ 4.5 5  4.51 -2  7 1  9 -1  9.01 4  10 3)
```

Notice that the element "line" really consists of two lines in the input file; the second line contains a plus sign (+) in the first column. SPICE interprets this to be a continuation of the element line immediately preceding. In order to approximate accurately a complex waveform using a piece-wise linear approximation, a large number of lines would have to be used. A PWL source element "line" describing an electrocardiogram signal every 5 ms for 500 ms uses 16 lines of the input file. In a case such as this, it would be wise to create a text file containing the complex PWL source by itself. Whenever that source was needed for a SPICE input file, the text file could be appended to the input file, saving a great deal of tedious typing.

3.5.1.5 Single-Frequency FM Function for Independent Sources

```
VXXXXXX    N+ N- SFFM(VO VA FC MDI FS)
IXXXXXX    N+ N- SFFM(VO VA FC MDI FS)
```

describe a voltage or current single-frequency FM (frequency modulation) source. The parameters are:

Parameter		Default Value	Unit
VO	(offset)	must specify	Volts or amps
VA	(amplitude)	must specify	Volts or amps
FC	(carrier frequency)	1/TSTOP	Hz
MDI	(modulation index)	must specify	none
FS	(signal frequency)	1/TSTOP	Hz

TSTOP refers to the step time in the transient analysis which is specified by a .TRAN TSTEP TSTOP control line in the input file. Figure 3.8 shows a single-frequency FM waveform illustrating VO and VA; it shows an unusually large amount of deviation (500 KHz with a carrier frequency of 1 MHz) and was created by the element line

VSAMPLE 1 0 SFFM(2 4 1MEG 5 100K)

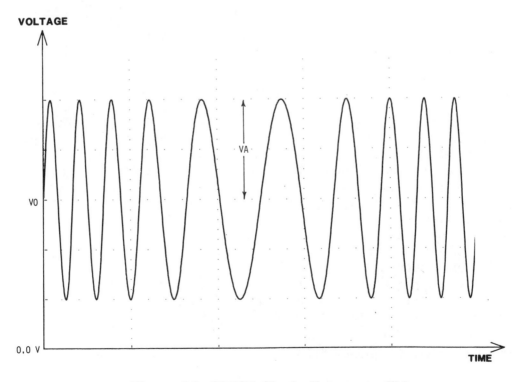

Figure 3.8 SPICE Single-Frequency FM Waveform.

The modulation index is the deviation (in Hz) divided by the modulating signal frequency (in Hz), and has no unit. The waveform as a function of time, t, is described by the relationship:

VXXXXXX = VO + VA*sine((2*PI*FC*t) + MDI*sine(2*PI*FS*t))

CHAPTER SUMMARY

SPICE is not an interactive program. You may have to submit an edited input file several times to get the results looking the way you want them.

Each node in a SPICE input file must have a DC path to node zero (ground). You may have to put in dummy resistors of very large value (so as not to disturb circuit operation) to satisfy this requirement.

Comments are very handy additions to SPICE input files.

Integer, floating point, scientific notation and engineering notation numbers are all usable in input files.

Element names can have up to 8 characters.

Independent current and voltage sources can have DC and AC values, and can be the following types of time-dependent functions: pulse, sinusoidal, exponential, piece-wise linear and single-frequency FM. It is best not to give a source DC, AC and time-dependent values, if only to avoid confusion. You can always run two or three different analyses, with the source edited to have only one value at a time.

Chapter 4

FIRST ATTEMPTS AT ANALYSIS WITH SAMPLE CIRCUITS

4.1   INTRODUCTION

In Chap. 3 some of the ground rules of using SPICE were covered.   In this chapter, a variety of examples will be presented to illustrate what an input file looks like and what the output file created by SPICE analysis contains.   By looking at three things, the circuit schematic diagram, the input file and the output file, you will gain an appreciation for what SPICE can do and you will start developing the ability to use SPICE yourself.   For a detailed explanation of each SPICE feature illustrated, refer to the chapter in which that feature is covered.   A good way to begin learning SPICE is to create input files similar to the ones in this chapter, and run analyses of them.

4.2   THE FIRST CIRCUIT – 3 RESISTORS, 1 BATTERY

Let's begin with a DC circuit, containing a battery (independent voltage source) and three resistors.   The circuit diagram is shown in Fig. 4.1.   A convenient node to choose for the datum node is the negative side of the battery; this is

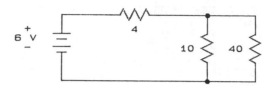

Figure 4.1 3 Resistor Circuit.

labeled node 0.   The two other nodes which must be labeled
are the positive side of the battery and the junction of the
three resistors.   Since the only restriction on node numbers
is that they be non-negative integers, let the two other
nodes be numbered 17 and 84 (any other positive integers
would do just fine!).   The notated circuit is shown in Fig.
4.2.

Any input file for SPICE analysis <u>must</u> contain the
following:

1. a title line

2. element lines for each element (R, L, C, etc.) in the
circuit

3. a .END line. A period (.) must appear before END.

Other lines, which are optional, include control lines
(which cause things to happen like changing the voltage or
frequency of a source, setting the temperature(s) of the circuit, specifying the type(s) of analyses to be done, printing
out a graph or table of results, etc.) and comment lines,
which make reading the input file much easier weeks or months
later.
Shown in Fig. 4.3, the input file for the circuit of Fig.
4.2 contains only a title line, four element lines and a .END
line.   Other than the title line and the .END line, the order

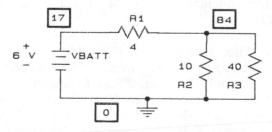

**Figure 4.2 3 Resistor Circuit Notated for
SPICE.**

```
3R.CIR  THREE RESISTOR, ONE BATTERY PROBLEM
VBATT   17  0  6
R1      17  84  4
R2      84  0  10
R3      84  0  40
.WIDTH OUT = 80
.END
```

**Figure 4.3 SPICE Input File for 3
Resistor Circuit.**

of all other lines is arbitrary. Notice that each element
has a name, as well as a value. For example, the 4 ohm resistor is named R1. The names are arbitrary as well, except
that all resistors must begin with an R, all voltage sources
must begin with a V, etc.

Although no specific analysis is requested, SPICE automatically does a DC analysis, called a "Small Signal Bias Solution," and prints the results in the output file. The title
line can contain anything whatsoever; however, in the interests of organization and preservation of one's sanity, it is
best to make the title contain two things:

1. the name of the input file. In this example, 3R.CIR
is the name of the file containing the title line, the
element lines and the .END line. Doing this makes it
easier to find an input file in the directory of your computer at a later time; there will be many input files
when you become a regular user of SPICE.

2. a brief description of what the circuit is (THREE RES-
ISTOR, ONE BATTERY PROBLEM). This is useful because both
the input file and the output file lack a schematic diagram of the circuit being analyzed. Though one can study
an input file and thereby reconstruct the circuit so described, it helps to have a clue in the title line, since
the title line normally will be printed on each page of
the output file.

The output file, shown in Fig. 4.4, contains two pages:
the first is a listing of the input file, called "Circuit Description," the second is the "Small Signal Bias Solution,"
which gives the DC voltage at each node other than the datum
(0) node. Notice that the voltage at node 17 is 6.000 VDC
(the battery voltage) and the voltage at node 84 is 4.000
VDC, both compared to the datum node. All node voltages produced by the Small Signal Bias Solution are with respect to

```
******* 1-DEC-87 ******* SPICE 2G.6    3/15/83 *******10:45:47*****

3R.CIR  THREE RESISTOR, ONE BATTERY PROBLEM

****    INPUT LISTING              TEMPERATURE =  27.000 DEG C


*********************************************************************
VBATT   17  0  6
R1      17 84  4
R2      84  0 10
R3      84  0 40
.WIDTH OUT = 80
.END

******* 1-DEC-87 ******* SPICE 2G.6    3/15/83 *******10:45:47*****

3R.CIR  THREE RESISTOR, ONE BATTERY PROBLEM

****    SMALL SIGNAL BIAS SOLUTION   TEMPERATURE =  27.000 DEG C


*********************************************************************
 NODE   VOLTAGE    NODE   VOLTAGE

( 17)   6.0000   ( 84)   4.0000


    VOLTAGE SOURCE CURRENTS

    NAME       CURRENT


    VBATT    -5.000D-01


    TOTAL POWER DISSIPATION  3.00D+00  WATTS


        JOB CONCLUDED

        TIME              PAGE    DIRECT   BUFFERED
    CPU         ELAPSED  FAULTS    I/O       I/O
 0: 0: 0.39  0: 0: 1.50    22       5         1

        TOTAL JOB TIME           0.38
```

**Figure 4.4 SPICE Output File for 3
Resistor Circuit.**

the datum node; thus, the voltage at node 0 is, by definition, zero volts.

4.3   HOW TO RUN SPICE
ON A MAINFRAME COMPUTER

In order to perform a SPICE analysis, the SPICE program must
be run on the particular computer system one is using. The
illustration below shows how it is run on a DEC VAX 11/780
mainframe computer (assuming that SPICE has been installed as
a system utility on the VAX). A similar procedure would be
used on another type of computer. The manager of the computer system should be able to tell you if SPICE is on your system, how to run SPICE and how to use the text editor.
     What you must type in is shown <u>underlined</u> in the steps
below. The steps involved are:

1. Create an input file describing the circuit and the desired analyses. Use a text editor (EDT on the VAX) to do
this. Call this input file FILENAME.CIR (FILENAME represents what you choose to name the file; 3R, DIFFAMP, RC-
AMP, etc. are typical names).

2. In DCL (Digital Command Language, the operating system
of the VAX), with the DCL $ prompt symbol showing, type
in <u>SPICE</u> and hit return.

3. SPICE will prompt you for the name of the input file
with INPUT FILE: . Type in <u>FILENAME.CIR</u> and hit
return. Remember, FILENAME represents the name of <u>your</u>
file.

4. SPICE will prompt you for the name you wish to call
the output file with OUTPUT FILE: . Type in
<u>FILENAME.OUT</u> and hit return.

5. SPICE will read the input file, check for errors, and
proceed to perform the analyses stated in the input
file. Results will be written to an output file called
FILENAME.OUT. You will be aware that the analysis is
over when the message FORTRAN STOP appears, followed by
the DCL $ prompt symbol.

6. To look at the output file, type in <u>TYPE FILENAME.OUT</u>
and hit return. The output file will appear on the CRT
monitor.

7. To obtain hardcopy of the output file, type in <u>PRINT FILENAME.OUT</u> and hit return. The output file will be printed on the VAX system line printer.

A sample run of SPICE on a VAX system, using the circuit (Fig. 4.1) and its input file 3R.CIR (Fig. 4.3), would look like this on the system video monitor:

```
$ SPICE
    INPUT FILE: 3R.CIR
    OUTPUT FILE: 3R.OUT
    FORTRAN STOP
$ TYPE 3R.OUT
$ PRINT 3R.OUT
```

One must remember that SPICE is not an interactive program; you have to look at the output file to see the results of the analysis. If the results are not what is desired, or if the analysis is to be run again with a parameter changed (e.g., substituting a different value of a resistor), a copy can be made of the original input file. Then the one or two lines in the copy of the input file can easily be edited to make the desired change, and the procedure in steps 2 through 7 would be followed again.

Appendix C lists other versions of SPICE, some of which can run on personal computers. These come with very detailed operating instructions, and will get you started doing circuit analysis on a PC.

In order to gain some proficiency in using SPICE, take a simple circuit (Fig. 4.1 or similar), notate it for SPICE analysis, write an input file describing the circuit, run SPICE, and look at the output file. If you make an error in the input file, SPICE will write meaningful error messages in the output file which will help you to find and correct the error.

4.4   HOW TO AVOID WASTING PAPER

When SPICE creates the output file, it puts a header at the top of each page and puts each part of the output file on a separate page. The header contains the date the analysis was done, the version of SPICE being used and its release date, the time the analysis was done, the title line from the input file, and a <u>lot</u> of asterisks (**). This can result in a very large amount of paper being used for certain types of analyses. While this may be useful for some applications, for the user who wants to look at the output file but conserve printer paper there is a simple fix.

One need only put the control line .OPTIONS NOPAGE (with no spaces in NOPAGE) in the input file, anywhere between the title line (which is always the first linc) and the .END line (which is always the last line). This control line will cause page ejects to be suppressed, concatenating the printout, and saving considerable amounts of paper. Figure 4.5 shows the same input file as Fig. 4.3, except that the control line .OPTIONS NOPAGE has been added. Note that this file was made by copying 3R.CIR to 3RNOPAGE.CIR and editing 3RNOPAGE.CIR. Editing here consisted of adding one control line and one comment line to the input file (and changing the title line).

The resulting output file (Fig. 4.6) is seen to have the same information as 3R.OUT, except that only one header appears and less paper is used.

4.5   THE SECOND CIRCUIT – THREE RESISTORS, TWO BATTERIES

In this DC circuit, there are two batteries. See Fig. 4.7. SPICE can be used to give a Small Signal Bias Solution, and thus determine the voltage at each node. In addition to solving the circuit, you might want to know what value would the 8 Volt source have to be so that the voltage across the 3K ohm resistor was 0? In order to solve this with SPICE, we shall make the battery on the right, the 8 Volt source, vary from 0 to 10 Volts, and do a DC analysis of the circuit (DC analysis is covered in depth in Chap. 6). The results of the

```
3RNOPAGE.CIR  THREE RESISTOR, ONE BATTERY PROBLEM
VBATT  17  0  6
R1     17  84  4
R2     84  0  10
R3     84  0  40
*     THE LINE BELOW SAVES PAPER
.OPTIONS NOPAGE
.WIDTH OUT = 80
.END
```

Figure 4.5 Input File 3RNOPAGE.CIR with .OPTIONS NOPAGE Added.

```
******* 1-DEC-87 *******  SPICE 2G.6    3/15/83 *******10:47:00*****


3RNOPAGE.CIR  THREE RESISTOR, ONE BATTERY PROBLEM


****     INPUT LISTING                    TEMPERATURE =   27.000 DEG C


*********************************************XXX*****************************



VBATT   17   0   6
R1      17  84   4
R2      84   0  10
R3      84   0  40
*    THE LINE BELOW SAVES PAPER
.OPTIONS NOPAGE
.WIDTH OUT = 80
.END

****     SMALL SIGNAL BIAS SOLUTION     TEMPERATURE =   27.000 DEG C

 NODE   VOLTAGE     NODE   VOLTAGE


( 17)   6.0000    ( 84)    4.0000




    VOLTAGE SOURCE CURRENTS

    NAME        CURRENT


    VBATT    -5.000D-01


    TOTAL POWER DISSIPATION   3.00D+00  WATTS


        JOB CONCLUDED


       TIME            PAGE    DIRECT   BUFFERED
    CPU        ELAPSED  FAULTS   I/O       I/O
 0: 0: 0.38  0: 0: 0.79    27      5         1


       TOTAL JOB TIME          0.37
```

Figure 4.6 Output File 3RNOPAGE.OUT.

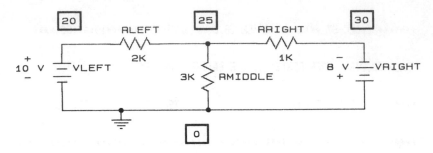

Figure 4.7 3 Resistor, 2 Battery Circuit.

analysis will be printed in tabular form and plotted as a graph of output voltage versus battery voltage. The input file, 3R2BAT.CIR, is shown in Fig. 4.8. The stepping of VRIGHT from 0 V to 10 V in steps of 0.5 V is accomplished by the control line .DC VRIGHT 0 10 0.5 . The printing of a table of the voltage at node 25 versus VRIGHT is done by the control line .PRINT DC V(25) . The control line .PLOT DC V(25) makes a graph of the DC voltage at node 25 versus VRIGHT.

The output file, 3R2BAT.OUT, is shown in Fig. 4.9. The first page is the Circuit Description, the next page is the table of node 25 voltage versus VRIGHT voltage, and the last page is a graph of node 25 voltage versus VRIGHT voltage. There are two columns of numbers on the left side of the graph: VRIGHT, the independent variable, and V(25), the dependent variable. The horizontal V(25) axis is labeled from -4 V to +4 V. An examination of the table or the graph shows that when VRIGHT is 8 V, the voltage at node 25 is -1.636 V; when VRIGHT is 5 V, the voltage at node 25 is 0 volts (shown as 3.661E-16 V, due to small calculation errors).

```
3R2BAT.CIR    VARY ONE BATTERY VOLTAGE
VLEFT      20  0  10
RLEFT      20 25  2K
RMIDDLE    25  0  3K
RRIGHT     25 30  1K
VRIGHT      0 30   8
.DC VRIGHT 0 10 0.5
*    THE LINE ABOVE SWEEPS VRIGHT FROM 0V TO 10V IN 0.5V STEPS
.PRINT DC V(25)
.PLOT  DC V(25)
.END
```

Figure 4.8 Input File 3R2BAT.CIR.

```
******** 9/ 2/87******** Demo PSpice (May 1986) *******15:55:56********


3R2BAT.CIR    VARY ONE BATTERY VOLTAGE


****    CIRCUIT DESCRIPTION


**********************************************************************



VLEFT      20  0  10
RLEFT      20 25  2K
RMIDDLE    25  0  3K
RRIGHT     25 30  1K
VRIGHT      0 30   8
.DC VRIGHT 0 10 0.5
*    THE LINE ABOVE SWEEPS VRIGHT FROM 0V TO 10V IN 0.5V STEPS
.PRINT DC V(25)
.PLOT  DC V(25)
.END



******** 9/ 2/87******** Demo PSpice (May 1986) *******15:55:56********


3R2BAT.CIR    VARY ONE BATTERY VOLTAGE


****    DC TRANSFER CURVES            TEMPERATURE =   27.000 DEG C


**********************************************************************


   VRIGHT        V(25)


    .000E+00     2.727E+00
   5.000E-01     2.455E+00
   1.000E+00     2.182E+00
   1.500E+00     1.909E+00
   2.000E+00     1.636E+00
   2.500E+00     1.364E+00
   3.000E+00     1.091E+00
   3.500E+00     8.182E-01
   4.000E+00     5.455E-01
   4.500E+00     2.727E-01
   5.000E+00     3.661E-16
   5.500E+00    -2.727E-01
```

Figure 4.9 Output File 3R2BAT.OUT.

```
6.000E+00      -5.455E-01
6.500E+00      -8.182E-01
7.000E+00      -1.091E+00
7.500E+00      -1.364E+00
8.000E+00      -1.636E+00
8.500E+00      -1.909E+00
9.000E+00      -2.182E+00
9.500E+00      -2.455E+00
1.000E+01      -2.727E+00
```

```
******** 9/ 2/87******** Demo PSpice (May 1986) *******15:55:56********

3R2BAT.CIR     VARY ONE BATTERY VOLTAGE

****    DC TRANSFER CURVES              TEMPERATURE =   27.000 DEG C

**************************************************************************

  VRIGHT       V(25)

                   -4.000D+00   -2.000D+00    .000D+00    2.000D+00    4.000D+00

     .000D+00  2.727D+00 .          .           .           .      *      .
    5.000D-01  2.455D+00 .          .           .           .    *        .
    1.000D+00  2.182D+00 .          .           .           .  .*         .
    1.500D+00  1.909D+00 .          .           .           *.            .
    2.000D+00  1.636D+00 .          .           .          *.             .
    2.500D+00  1.364D+00 .          .           .       *  .              .
    3.000D+00  1.091D+00 .          .           .     *    .              .
    3.500D+00  8.182D-01 .          .           .    *     .              .
    4.000D+00  5.455D-01 .          .           .   *      .              .
    4.500D+00  2.727D-01 .          .           . *        .              .
    5.000D+00  3.661D-16 .          .           *          .              .
    5.500D+00 -2.727D-01 .          .        * .           .              .
    6.000D+00 -5.455D-01 .          .       * .            .              .
    6.500D+00 -8.182D-01 .          .      *  .            .              .
    7.000D+00 -1.091D+00 .          .   *     .            .              .
    7.500D+00 -1.364D+00 .          . *       .            .              .
    8.000D+00 -1.636D+00 .        . *         .            .              .
    8.500D+00 -1.909D+00 .        .*          .            .              .
    9.000D+00 -2.182D+00 .       *.           .            .              .
    9.500D+00 -2.455D+00 .      * .           .            .              .
    1.000D+01 -2.727D+00 .     *  .           .            .              .

        JOB CONCLUDED

        TOTAL JOB TIME          5.10
```

Figure 4.9 (Continued).

4.6   THE THIRD CIRCUIT – PARALLEL RESONANT TANK, AC ANALYSIS

In the schematic diagram, Fig. 4.10, a 1 V sinusoidal voltage source in series with a 1K ohm resistor is connected across a lossy tank circuit. The resonant frequency of this low-Q tank L-C is about 10 KHz. The SPICE input file in Fig. 4.11 contains the 5 element lines for the circuit, and has a control line which causes the frequency of the voltage source to vary from 5 KHz to 15 KHz, in 20 linearly-spaced steps. This causes the AC analysis to be done at 21 frequencies -- every 500 Hz. The results of the AC analysis will appear in the output file because of the control lines .PRINT AC VM(2) VP(2) and .PLOT AC VM(2) VP(2) . The .PRINT command causes a table to be printed, with columns of frequency, voltage magnitude at node 2, and voltage phase at node 2. The .PLOT command causes a graph of voltage magnitude at node 2 and voltage phase at node 2 to be plotted versus frequency.

The output file, Fig. 4.12, contains the circuit description, a small-signal bias solution (which shows DC values to be zero, since there are no DC sources in the circuit), a table of results, and a graph of results. The magnitude of the node 2 voltage is seen to peak at 10 KHz (labeled 1.000E+04 by SPICE), at 568.7 mV. The phase of the voltage at node 2 is 80.35 degrees at 5 KHz, -1.173 degrees at 10 KHz, and -84.25 degrees at 15 KHz; these data show that resonance occurs quite near 10 KHz. Although SPICE can plot up to 8 parameters on one graph, only the first parameter listed in the .PLOT control line will have its data printed as well as plot-

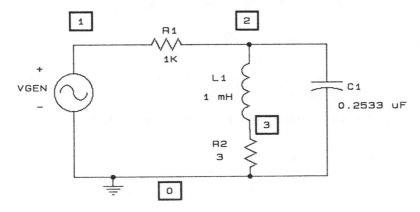

Figure 4.10   Tank Circuit.

```
ACTANK.CIR   PARALLEL RESONANT CIRCUIT, 10 KHZ
R1  1  2  1K
R2  3  0  3
VGEN 1 0 AC 1
L1  2  3  1M
C1  2  0  0.2533U
.AC LIN 21  5000 15000
.PRINT AC VM(2) VP(2)
.PLOT AC  VM(2) VP(2)
.END
```

Figure 4.11 Input File ACTANK.CIR.

```
******** 9/ 9/87******** Demo PSpice (May 1986) *******10:14:51********


ACTANK.CIR   PARALLEL RESONANT CIRCUIT, 10 KHZ


****     CIRCUIT DESCRIPTION


*************************************************************************



R1  1  2  1K
R2  3  0  3
VGEN 1 0 AC 1
L1  2  3  1M
C1  2  0  0.2533U
.AC LIN 21  5000 15000
.PRINT AC VM(2) VP(2)
.PLOT AC  VM(2) VP(2)
.END



******** 9/ 9/87******** Demo PSpice (May 1986) *******10:14:51********


ACTANK.CIR   PARALLEL RESONANT CIRCUIT, 10 KHZ


****     SMALL SIGNAL BIAS SOLUTION      TEMPERATURE =   27.000 DEG C


*************************************************************************


NODE    VOLTAGE    NODE    VOLTAGE    NODE    VOLTAGE


(  1)     .0000    (  2)     .0000    (  3)     .0000
```

Figure 4.12 Output File ACTANK.OUT.

```
******** 9/ 9/87******** Demo PSpice (May 1986) *******10:14:51********

ACTANK.CIR   PARALLEL RESONANT CIRCUIT, 10 KHZ

****    AC ANALYSIS                      TEMPERATURE =   27.000 DEG C


**********************************************************************


     FREQ       VM(2)       VP(2)


     5.000E+03   4.180E-02   8.035E+01
     5.500E+03   4.933E-02   8.008E+01
     6.000E+03   5.850E-02   7.956E+01
     6.500E+03   7.001E-02   7.874E+01
     7.000E+03   8.497E-02   7.752E+01
     7.500E+03   1.053E-01   7.570E+01
     8.000E+03   1.348E-01   7.289E+01
     8.500E+03   1.808E-01   6.826E+01
     9.000E+03   2.608E-01   5.969E+01
     9.500E+03   4.111E-01   4.086E+01
     1.000E+04   5.687E-01  -1.173E+00
     1.050E+04   4.319E-01  -4.320E+01
     1.100E+04   2.882E-01  -6.204E+01
     1.150E+04   2.103E-01  -7.067E+01
     1.200E+04   1.652E-01  -7.539E+01
     1.250E+04   1.364E-01  -7.831E+01
     1.300E+04   1.165E-01  -8.028E+01
     1.350E+04   1.019E-01  -8.170E+01
     1.400E+04   9.079E-02  -8.276E+01
     1.450E+04   8.204E-02  -8.359E+01
     1.500E+04   7.496E-02  -8.425E+01
```

```
******** 9/ 9/87******** Demo PSpice (May 1986) *******10:14:51********

ACTANK.CIR   PARALLEL RESONANT CIRCUIT, 10 KHZ

****    AC ANALYSIS                      TEMPERATURE =   27.000 DEG C


**********************************************************************
```

Figure 4.12 (Continued).

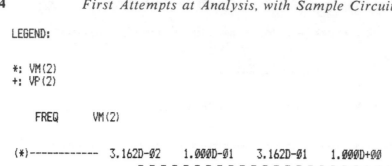

```
LEGEND:

*: VM(2)
+: VP(2)


   FREQ      VM(2)


(*)------------  3.162D-02   1.000D-01   3.162D-01   1.000D+00   3.162D+00
                 - - - - - - - - - - - - - - - - - - - - - - - - - - - -

(+)------------ -1.000D+02  -5.000D+01    .000D+00   5.000D+01   1.000D+02
                 - - - - - - - - - - - - - - - - - - - - - - - - - - - -
```

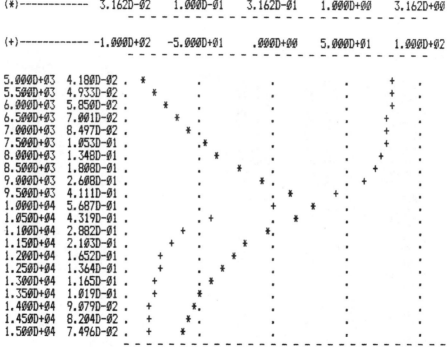

```
5.000D+03  4.180D-02 .  *           .           .           .    +     .
5.500D+03  4.933D-02 .    *         .           .           .    +     .
6.000D+03  5.850D-02 .      *       .           .           .    +     .
6.500D+03  7.001D-02 .       *      .           .           .   +      .
7.000D+03  8.497D-02 .         *    .           .           .   +      .
7.500D+03  1.053D-01 .           .* .           .           .   +      .
8.000D+03  1.348D-01 .           .    *         .           .  +       .
8.500D+03  1.808D-01 .           .        *     .           . +        .
9.000D+03  2.608D-01 .           .           *  .           .+         .
9.500D+03  4.111D-01 .           .           .  *   .      + .         .
1.000D+04  5.687D-01 .           .           .    + .    *   .         .
1.050D+04  4.319D-01 .           .     .  +   .    *       .           .
1.100D+04  2.882D-01 .           .   +     .       *.          .        .
1.150D+04  2.103D-01 .           .  +      .      *  .          .        .
1.200D+04  1.652D-01 .        +   .        .    *    .          .        .
1.250D+04  1.364D-01 .        +   .        .  *      .          .        .
1.300D+04  1.165D-01 .      +     .        . *       .          .        .
1.350D+04  1.019D-01 .      +     .    *   .         .          .        .
1.400D+04  9.079D-02 .    +       .  *.    .         .          .        .
1.450D+04  8.204D-02 .    +     *.  .      .         .          .        .
1.500D+04  7.496D-02 .    +     *   .      .         .          .        .
                     - - - - - - - - - - - - - - - - - - - - - - - - - -
```

```
JOB CONCLUDED

TOTAL JOB TIME          5.90
```

Figure 4.12 (Continued).

ted on the graph. Plotting more than 3 parameters on the
same graph is recommended only for those who love puzzles and
have excellent vision. You can use as many .PLOT control
lines as needed, so it is never necessary to put more than 3
plots on one graph.

4.7   THE FOURTH CIRCUIT – RC CIRCUIT WITH PULSE INPUT

In this example, an R-C circuit will be connected to a pulse voltage source, and an analysis of circuit voltage versus time, called a transient analysis, will be performed. The result of the transient analysis will appear in the output file in tabular form and as a graph of voltage versus time. The circuit is shown in Fig. 4.13. The single pulse input source goes from 0 V to 1 V after 10 microseconds of delay (relative to the beginning of the transient analysis), with rise and fall times of 1 nanosecond, and has a duration of 80 microseconds. It is described in the SPICE input file by the element line   VIN 1 0 PULSE(0 1 10U 1N 1N 80U) . The results of the transient analysis will appear in the output file because of the control lines .PRINT TRAN V(2) V(1) and .PLOT TRAN V(2) V(1) . The input file is shown in Fig. 4.14.

Since there are no DC sources in the circuit, and the pulse level is 0 V at time zero, the output file (Fig. 4.15) indicates that at time zero the voltage at nodes 1 and 2 is 0 V. The table of transient analysis results and the graph of V(1) and V(2) versus time show that the input voltage, V(1), is 0 V up through 10 usec, is 1 V from 15 to 90 usec and returns to 0 V at 95us where it remains until 160 usec. The voltage across the capacitor, V(2), begins rising exponentially at 10 usec, reaches 981.8 mV at 90 usec, and exponentially decays to 29.44 mV at 160 usec. The input voltage and the capacitor voltage are identical or nearly so at 0, 5, 10 and 90 usec. This overlap of the two graphs is indicated by an X symbol at those four times.

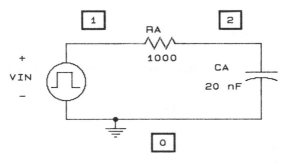

Figure 4.13 R-C Circuit with Pulse Input.

```
PULSE.CIR   R-C CIRCUIT WITH PULSE INPUT
VIN 1 Ø PULSE(Ø 1 1ØU 1N 1N 8ØU)
RA  1  2  1ØØØ
CA  2  Ø  2ØN
.TRAN 5U 16ØU
.PRINT TRAN V(2) V(1)
.PLOT  TRAN V(2) V(1)
.END
```

Figure 4.14 Input File PULSE.CIR.

```
******** 9/ 9/87******** Demo PSpice (May 1986) ******* 9:47:16********


PULSE.CIR   R-C CIRCUIT WITH PULSE INPUT


****    CIRCUIT DESCRIPTION


***************************************************************************


VIN 1 Ø PULSE(Ø 1 1ØU 1N 1N 8ØU)
RA  1  2  1ØØØ
CA  2  Ø  2ØN
.TRAN 5U 16ØU
.PRINT TRAN V(2) V(1)
.PLOT  TRAN V(2) V(1)
.END




******** 9/ 9/87******** Demo PSpice (May 1986) ******* 9:47:16********


PULSE.CIR   R-C CIRCUIT WITH PULSE INPUT


****    INITIAL TRANSIENT SOLUTION      TEMPERATURE =   27.ØØØ DEG C


***************************************************************************

  NODE   VOLTAGE     NODE   VOLTAGE

 (  1)     .ØØØØ   (  2)     .ØØØØ
```

Figure 4.15 Output File PULSE.OUT.

```
******** 9/ 9/87******** Demo PSpice (May 1986) ******* 9:47:16********

PULSE.CIR   R-C CIRCUIT WITH PULSE INPUT

****      TRANSIENT ANALYSIS              TEMPERATURE =   27.000 DEG C

*************************************************************************


      TIME        V(2)        V(1)

     .000E+00    .000E+00    .000E+00
    5.000E-06    .000E+00    .000E+00
    1.000E-05    .000E+00    .000E+00
    1.500E-05   2.189E-01   1.000E+00
    2.000E-05   3.932E-01   1.000E+00
    2.500E-05   5.269E-01   1.000E+00
    3.000E-05   6.320E-01   1.000E+00
    3.500E-05   7.135E-01   1.000E+00
    4.000E-05   7.769E-01   1.000E+00
    4.500E-05   8.266E-01   1.000E+00
    5.000E-05   8.648E-01   1.000E+00
    5.500E-05   8.950E-01   1.000E+00
    6.000E-05   9.181E-01   1.000E+00
    6.500E-05   9.363E-01   1.000E+00
    7.000E-05   9.504E-01   1.000E+00
    7.500E-05   9.614E-01   1.000E+00
    8.000E-05   9.700E-01   1.000E+00
    8.500E-05   9.766E-01   1.000E+00
    9.000E-05   9.818E-01   1.000E+00
    9.500E-05   7.651E-01    .000E+00
    1.000E-04   5.970E-01    .000E+00
    1.050E-04   4.635E-01    .000E+00
    1.100E-04   3.617E-01    .000E+00
    1.150E-04   2.812E-01    .000E+00
    1.200E-04   2.190E-01    .000E+00
    1.250E-04   1.705E-01    .000E+00
    1.300E-04   1.326E-01    .000E+00
    1.350E-04   1.034E-01    .000E+00
    1.400E-04   8.021E-02    .000E+00
    1.450E-04   6.262E-02    .000E+00
    1.500E-04   4.865E-02    .000E+00
    1.550E-04   3.793E-02    .000E+00
    1.600E-04   2.944E-02    .000E+00


******** 9/ 9/87******** Demo PSpice (May 1986) ******* 9:47:16********

PULSE.CIR   R-C CIRCUIT WITH PULSE INPUT

****      TRANSIENT ANALYSIS              TEMPERATURE =   27.000 DEG C
```

Figure 4.15 (Continued).

****

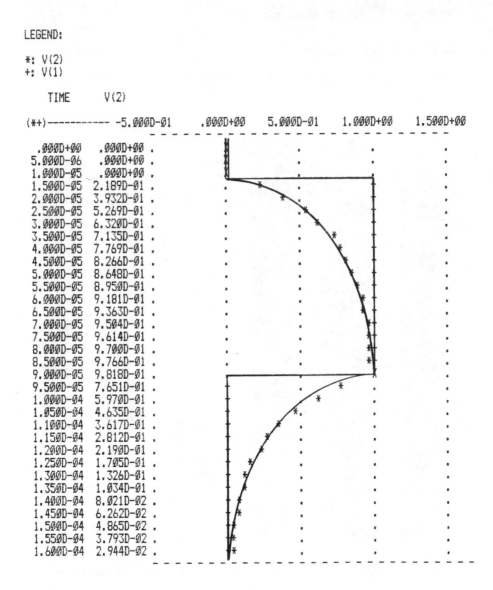

```
LEGEND:

*: V(2)
+: V(1)

     TIME      V(2)

 (*+)------------ -5.000D-01     .000D+00    5.000D-01    1.000D+00    1.500D+00
    .000D+00    .000D+00 .                    .            .            .
   5.000D-06    .000D+00 .                    .            .            .
   1.000D-05    .000D+00 .                    .            .            .
   1.500D-05   2.189D-01 .                    .            .            .
   2.000D-05   3.932D-01 .                    .            .            .
   2.500D-05   5.269D-01 .                    .            .            .
   3.000D-05   6.320D-01 .                    .            .            .
   3.500D-05   7.135D-01 .                    .            .            .
   4.000D-05   7.769D-01 .                    .            .            .
   4.500D-05   8.266D-01 .                    .            .            .
   5.000D-05   8.648D-01 .                    .            .            .
   5.500D-05   8.950D-01 .                    .            .            .
   6.000D-05   9.181D-01 .                    .            .            .
   6.500D-05   9.363D-01 .                    .            .            .
   7.000D-05   9.504D-01 .                    .            .            .
   7.500D-05   9.614D-01 .                    .            .            .
   8.000D-05   9.700D-01 .                    .            .            .
   8.500D-05   9.766D-01 .                    .            .            .
   9.000D-05   9.818D-01 .                    .            .            .
   9.500D-05   7.651D-01 .                    .            .            .
   1.000D-04   5.970D-01 .                    .            .            .
   1.050D-04   4.635D-01 .                    .            .            .
   1.100D-04   3.617D-01 .                    .            .            .
   1.150D-04   2.812D-01 .                    .            .            .
   1.200D-04   2.190D-01 .                    .            .            .
   1.250D-04   1.705D-01 .                    .            .            .
   1.300D-04   1.326D-01 .                    .            .            .
   1.350D-04   1.034D-01 .                    .            .            .
   1.400D-04   8.021D-02 .                    .            .            .
   1.450D-04   6.262D-02 .                    .            .            .
   1.500D-04   4.865D-02 .                    .            .            .
   1.550D-04   3.793D-02 .                    .            .            .
   1.600D-04   2.944D-02 .                    .            .            .


              JOB CONCLUDED

              TOTAL JOB TIME        12.30
```

Figure 4.15 (Continued).

CHAPTER SUMMARY

Four examples of using SPICE for circuit analysis have been presented in this chapter:

1. A DC circuit with one battery, without any control lines in the input file. SPICE by default performed a "Small Signal Bias Solution" which gave the node voltages. A slight change to the input file showed the paper-saving effects of adding a .OPTIONS NOPAGE control line. See input files 3R.CIR and 3RNOPAGE.CIR.

2. A DC circuit with two batteries; one battery was stepped through a range of voltages to illustrate performing a DC analysis. The results were plotted and presented in tabular form. See input file 3R2BAT.CIR.

3. A parallel resonant (tank) circuit was stepped over a range of frequencies, and an AC analysis was performed at each frequency for the voltage across the tank. The results were presented in tabular form and plotted. See input file ACTANK.CIR.

4. A series resistor-capacitor was connected to an input voltage pulse. A transient analysis was done, and the exponentially varying capacitor voltage was shown in tabular form and plotted. See input file PULSE.CIR.

By looking at these examples, and modifying them as suggested in the problems section which follows, you will gain proficiency at "talking" in SPICE language and using it for problems of your own.

PROBLEMS

1. Refer to the circuit schematic diagram in Fig. 4.16, which has nodes notated for SPICE analysis.
   a. Write a SPICE input file describing the circuit, including a title line and a .END line.
   b. Analyze the circuit for the DC voltage at node 12 (compared to ground), using conventional circuit analysis methods.
   c. Run a SPICE analysis using the input file from Prob. 1.a. How does your answer in 1.b compare with the SPICE analysis result?

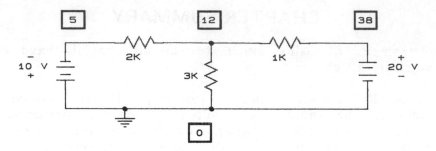

Figure 4.16 Circuit for Prob. 1.

2. Refer to the circuit diagram in Fig. 4.10.
a. Create an input file for that circuit, but replace the 3 ohm resistor with a 1 ohm resistor.
b. Predict what will happen to the shape of the graph of voltage magnitude at node 2 versus frequency. (Hint: a smaller resistor will affect the circuit Q, or quality factor).
c. Run a SPICE analysis using the input file from 2.a. Compare the graph that results with your prediction from 2.b.

3. In Fig. 4.13, replace the 20 nF capacitor with a 20 mH inductor, and repeat the SPICE analysis.

4. Use SPICE to find the voltage at node 16 in Fig. 4.17.

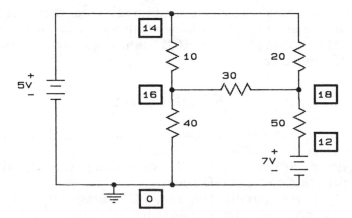

Figure 4.17 Circuit for Prob. 4.

Chapter 5

LINEAR DEPENDENT SOURCES

5.1 INTRODUCTION

SPICE allows the user to use dependent (or controlled) voltage sources and current sources in describing circuits. This can be useful in making simple models of semiconductor devices such as bipolar junction transistors, field-effect transistors and operational amplifiers. In this chapter linear controlled sources will be introduced; Appendix C will present the topic of nonlinear dependent sources.

The four kinds of linear dependent sources are:

> voltage-controlled current source, or VCCS
> voltage-controlled voltage source, or VCVS
> current-controlled current source, or CCCS
> current-controlled voltage source, or CCVS

The sources are uniquely specified by an element name beginning with the letter G, E, F or H, and have units as described below:

ELEMENT	EQUATION	ELEMENT TYPE	UNIT
VCCS	$I = G(V)$	Transconductance	Siemen
VCVS	$V = E(V)$	Voltage gain	V/V
CCCS	$I = F(I)$	Current gain	A/A
CCVS	$V = H(I)$	Transresistance	Ohm

5.2   LINEAR VCCS

Linear voltage-controlled current sources are described by the following element line

    GXXXXXX   N+ N- NC+ NC- VALUE

where N+ and N- are the nodes to which the current source is connected,   NC+ and NC- are the nodes which define the controlling voltage, and VALUE is the transconductance of the VCCS in siemens.

EXAMPLE:

    GWHIZ   4 13 22 26   5E-2
        The current source connected between nodes 4 and 13 (with positive current assumed to flow from 4 to 13) is controlled by the voltage difference between nodes 22 and 26 (node 22 considered positive).   The transconductance is 50 millisiemens. See Fig. 5.1.

In order to illustrate the use of the VCCS element, a simple model of a JFET will be done with a VCCS.   The circuit in Fig. 5.2, with nodes notated for SPICE analysis, can be redrawn with a VCCS replacing the JFET.   See Fig. 5.3.   The element lines describing the redrawn circuit are shown below. The transconductance of the JFET and the VCCS is assumed to be 20 mS.

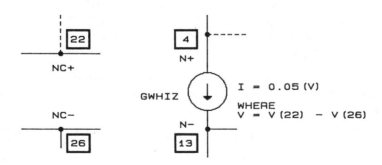

Figure 5.1 Voltage-Controlled Current Source.

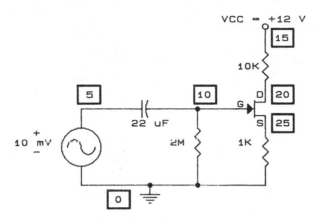

Figure 5.2   JFET Circuit Notated for SPICE.

VCC	15	0	12		
VSIGNAL	5	0	AC	10M	
CGATE	5	10	22U		
RGATE	10	0	2MEG		
RSOURCE	25	0	1K		
RDRAIN	15	20	10K		
GFET	20	25	10	25	20E-3

In this JFET modeling example, the current that flows from drain to source is determined by the gate-source voltage, V_{GS}, and the JFET transconductance of 20 mS.

$$I_{DS} = .020 \, (V_{GS})$$

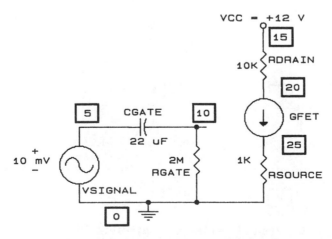

Figure 5.3   JFET Circuit Modeled with VCCS.

5.3   LINEAR VCVS

Linear voltage-controlled voltage sources are described by the following element line

EXXXXXX   N+ N- NC+ NC- VALUE

where N+ and N- are the nodes to which the voltage source is connected,   NC+ and NC- are the nodes which define the controlling voltage, and VALUE is the voltage gain of the VCVS in V/V.

EXAMPLE:

EGAIN   14 9 6 2   250
   The voltage source connected between nodes 14 and 9 (node 14 is positive) has a magnitude of 250 times the voltage between nodes 6 and 2. Refer to Fig. 5.4.

A good example of the use of the VCVS element is as a small-signal model of an operational amplifier at low frequencies (where slew-rate and gain-bandwidth limitations can be neglected).   The circuit in Fig. 5.5, with nodes notated for SPICE analysis, can be redrawn with an input resistance and a VCVS replacing the op-amp.   The operational amplifier has a differential input resistance of 1 megohm and an open-loop gain of 50,000 V/V.   See Fig. 5.6.   The element lines describing the redrawn circuit are shown below.

VSIG   4   0   AC   1
RSIG   4   11   50
RFB   22   19   20K

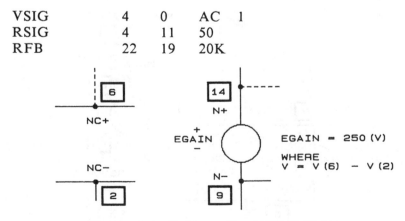

Figure 5.4 Voltage-Controlled Voltage Source.

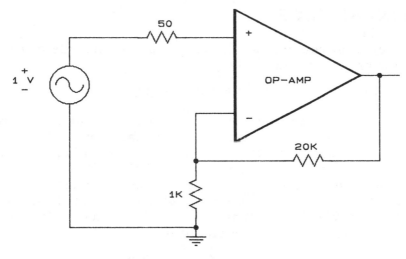

Figure 5.5 Op-Amp Circuit.

R1	19	0	1K	
RINOPAMP	11	19	1MEG	
EOPAMP	22	0	11	19	5E4

In this op-amp model, the voltage at the output of the op-amp is equal to the input differential voltage (voltage at node 11 minus the voltage at node 19) multiplied by the open-loop gain of 50,000 V/V.   Of course, this simple model neglects saturation, capacitance and slew-rate, among others.

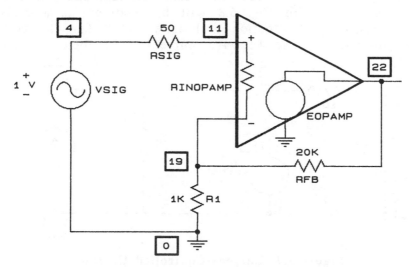

Figure 5.6 Op-Amp Circuit Modeled with VCVS.

5.4 LINEAR CCCS

Linear current-controlled current sources are described by the following element line

FXXXXXX   N+ N- VNAME VALUE

where N+ and N- are the nodes to which the current source is connected, VNAME is the name of a voltage source in the same input file through which the current controlling the CCCS flows, and VALUE is the current gain of the CCCS in A/A. Note that it may be necessary to insert a dead voltage source in the circuit branch where the controlling current flows in order to have a VNAME to control the CCCS. Such a dead voltage source would have no effect on circuit operation.

EXAMPLE:

FOUT 37 24 VSENSE 5
        Refer to Fig. 5.7. The current in CCCS FOUT connected between nodes 37 and 24 has a magnitude of 5 times the current flowing (downward from node 6 to node 11) through voltage source VSENSE and the 20 ohm resistor. Positive current in FOUT flows from node 37 to node 24, through the current source.

Bipolar junction transistors can be thought of as current-controlled current sources for small-signal operation. The BJT circuit shown in Fig. 5.8 will be modeled as a resistance for the base-emitter junction and as a CCCS for the collector-emitter junction. The forward current gain, beta, is assumed

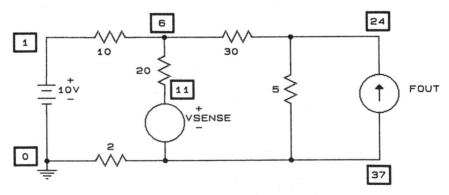

Figure 5.7 Current-Controlled Current Source Circuit.

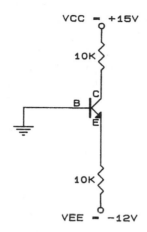

Figure 5.8   Bipolar Junction Transistor Circuit.

to be 80 A/A.   Figure 5.9 shows the modeled circuit.   The input element lines for Fig. 5.9 are shown below.

VIB	0	2	0
FBJT	4	5	VIB 80
VCC	3	0	15
VEE	6	0	-12
RBASE	2	5	1.8K
RE	5	6	10K
RC	3	4	10K

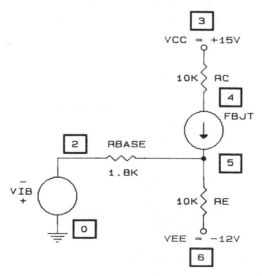

Figure 5.9   BJT Circuit Modeled with CCCS.

The resistance between node 2 and 5 (RBASE) is the hybrid parameter (hie), for which 1.8K ohms is appropriate for this circuit. Note that the dummy voltage source VIB (its magnitude is zero) was added to the circuit in order to provide a way of measuring the base current flowing up from ground into the base terminal of the model (node 2). VIB's positive node is 0 and its negative node is 2; this way current flowing from node 0 to node 2 is considered positive. While this is not a very practical model of a BJT, it does illustrate the format for a CCCS element line.

5.5   LINEAR CCVS

Linear current-controlled voltage sources are described by the following element line

HXXXXXX   N+ N- VNAME VALUE

where N+ and N- are the nodes to which the voltage source is connected, VNAME is the name of a voltage source in the same input file through which the current controlling the CCVS flows, and VALUE is the transresistance of the CCVS in ohms. Note that it may be necessary to insert a dead voltage source in the circuit branch where the controlling current flows, in order to have a VNAME to control the CCVS. Such a dead voltage source would have no effect on circuit operation.

EXAMPLE:

HSAMPLE   24 37 VSENSE 6
Refer to Fig. 5.10. The voltage across CCVS HSAMPLE, connected between nodes 24 and 37, has a magnitude of 6 times the current flowing (downward from node 6 to node 11) through voltage source VSENSE and the 20 ohm resistor.

This chapter has been devoted to presenting linear dependent sources. SPICE is capable of analyzing circuits containing non-linear dependent sources as well. Non-linear dependent sources are discussed in Appendix B.

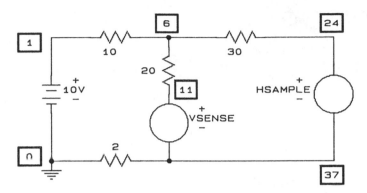

Figure 5.10 Current-Controlled Voltage Source Circuit.

CHAPTER SUMMARY

SPICE has four kinds of linear dependent sources: VCCS, VCVS, CCCS and CCVS.

Linear dependent sources can be used to create your own models of circuit devices.

All four kinds of linear dependent sources are connected to two nodes. Two kinds (VCCS and VCVS) are controlled by the difference in voltage between two nodes, while the other two kinds (CCCS and CCVS) are controlled by the current through an independent voltage source. This independent voltage source (which may be a dead source) may have to be added to a branch to sense the current in that branch.

Chapter 6

TYPES OF ANALYSES

6.1   INTRODUCTION

By including control lines in an input file, SPICE can be made to perform many kinds of analyses of a circuit. Generally, only one or two will be done at a time. Later in this chapter use of each control line will be illustrated with examples. The many examples in Chap. 12, including circuit schematic diagrams, input files, output files and a discussion of each example will be helpful as well. The control lines and a brief description of the types of analysis are:

.DC   A non-linear analysis which determines the DC operating point of the circuit. Capacitors are open-circuited, and inductors are short-circuited during DC analysis. One or more sources (current and voltage) may be stepped over a range. Results of the analysis at each source value may be printed in a table and/or plotted.

.AC   This is a linear small-signal analysis. SPICE determines the DC operating point of a circuit and thereby calculates values for small-signal models of all nonlinear devices (semiconductors, inductors, capacitors, polynomial dependent sources).   Then the linear small-signal circuit model is analyzed at each frequency specified by the user.   Results of the analysis at each frequency can be printed and/or plotted.

.TRAN   This transient analysis determines the output variables (current and voltage) as a function of time over a time interval which is specified.   The size of the time step used within the interval may be specified.   A table can be printed and/or a plot made of the output variables versus time.

.OP   Causes SPICE to solve for and print the DC operating point of a circuit.   This is done automatically when an AC analysis is performed, to find the AC small-signal model parameters, and also when a transient analysis is done, to determine initial conditions.

.TF   Does a small-signal DC transfer function analysis of a circuit from a specified input to a specified output. The input resistance, output resistance, and transfer function (voltage gain, current gain, transresistance or transconductance) are printed.

.SENS   SPICE finds the DC small-signal sensitivities of one or more specified output variables with respect to every parameter in the circuit, including semiconductor parameters.   A large amount of paper can be consumed printing the results of this analysis for any but the smallest circuits !

.DISTO   If this control line is present with a .AC control line, SPICE will determine several kinds of distortion of the circuit in small-signal analysis.   Results may be printed and/or plotted.

.NOISE   When present with a .AC control line, SPICE will find the equivalent output noise and equivalent input noise at specified output and input points in the circuit. Results may be printed and/or plotted.

.FOUR   Performs a Fourier analysis of an output variable when done with a transient analysis.   Computes the amplitudes and phases of the first nine frequency components (harmonics) of a user-specified fundamental frequency, as well as the DC component.

.TEMP   Not truly an analysis by itself, .TEMP tells SPICE at what temperature(s) to simulate a circuit.

6.2   DC ANALYSIS USING .DC CONTROL LINE

Control line format:

.DC SRC START STOP INCR <SRC2 START2 STOP2 INCR2>

In the control line above, SRC is the name of the independent voltage or current source being varied.   START is the starting value, STOP is the final value, and INCR is the increment size;   units are volts or amps.   Optionally, a second source (SRC2) may be specified, along with its associated parameters.   If it is, the first source will be swept over its entire range (START to STOP) for each value of the second source.   Associated with any .DC control line should be a .PRINT or a .PLOT control line which will cause an output to be written to the SPICE output file.   For more information on .PRINT and .PLOT, see Chap. 7.

EXAMPLES:

.DC VGATE -2 3 0.5
    Independent voltage source VGATE (specified elsewhere in the input file) is incremented from -2 V to +3 V in 0.5 V steps.   In other words, it will be set to values of -2, -1.5, -1, . . . , +3 V.

.DC IBASE 0 100U 10U VCE 0 20 4
    Independent current source IBASE will step from 0 to 100 uA in steps of 10 uA while independent voltage source VCE is fixed at 0 V.   Then VCE will jump to 4 V, and IBASE will step again from 0 to 100 uA.   This process will be repeated until VCE reaches 20 V.   A total of 66 DC analyses will occur; 11 for IBASE, multiplied by 6 for VCE.

6.3   AC ANALYSIS USING .AC CONTROL LINE

Control line format:

    .AC LIN NP FSTART FSTOP
    .AC DEC ND FSTART FSTOP
    .AC OCT NO FSTART FSTOP

In the control lines above, FSTART is the lowest frequency (which may not be negative or zero), and FSTOP is the highest frequency.   The frequencies at which the analyses are done can be specified three ways:

> LIN   NP   will cause an analysis to occur at a number of frequencies equal to NP, linearly spaced between FSTART and FSTOP.   To do an AC analysis at a single frequency, set NP equal to 1, and make FSTART and FSTOP equal the desired frequency.

> DEC   ND   will make SPICE divide the frequency range into decades, with ND frequencies per decade.   Frequencies will be logarithmically-spaced (the ratio of any two adjacent frequencies is the same over the entire range). This decade spacing is most useful with wide ranges of frequency; if FSTOP is not an integer number of decades above FSTART, the highest frequency can exceed FSTOP.

> OCT   NO   instructs SPICE to break the frequency range into octaves, with NO frequencies per octave.   Frequencies will be logarithmically-spaced (the ratio of any two adjacent frequencies is the same over the entire range). This octave spacing is most useful with wide ranges of frequency; if FSTOP is not an integer number of octaves above FSTART, the highest frequency can exceed FSTOP.

Unlike the .DC control line, the .AC control line does not specify which source(s) are to be swept in frequency.   The frequency of all AC sources in the circuit will be set simultaneously by the .AC control line.   Associated with any .AC control line should be a .PRINT or a .PLOT control line which will cause an output to be written to the SPICE output file. For more information on .PRINT and .PLOT, see Chap. 7.

EXAMPLES:

.AC LIN 101 5000 6000
    Any independent AC sources in the circuit will start at 5 KHz and increment upward by 10 Hz until 6 KHz is reached. Thus, AC analysis will occur at frequencies of 5000, 5010, 5020, . . . , 5990, 6000 Hz.

.AC LIN 1 3.58MEG 3.58MEG
    All independent AC sources are set to 3.58 MHz for the AC analysis.

.AC DEC 5 1 10K

Since there are 4 decades of frequency [log(10K/1) = 4], and 5 points per decade, AC analysis will occur at 21 frequencies logarithmically-spaced between 1 Hz and 10 KHz.

.AC OCT 7 10 5120

The number of octaves of frequency can be determined by: NO = [(log(5120/10))/log2] = 9. AC analysis will be done at 64 frequencies (7 points/octave, 9 octaves) spaced logarithmically between 10 Hz and 5.12 KHz.

6.4   TRANSIENT ANALYSIS USING .TRAN CONTROL LINE

Control line format:

.TRAN TSTEP TSTOP <TSTART <TMAX>> <UIC>

In the control line above, TSTEP is the time increment used for plotting and/or printing results of the transient analysis; it is not necessarily the computing time step that SPICE uses between successive analyses. TSTOP is the time of the last transient analysis. TSTART may be omitted; its default value is zero. SPICE always starts the analysis at time zero; if a TSTART value greater than zero is specified, SPICE will analyze the circuit from zero to TSTART but will not store the analysis results.

The largest computing time step SPICE will use is TMAX. If TMAX is not specified, the computing time step will be the smaller of TSTEP or (TSTOP - TSTART)/50. TMAX can be used to ensure that the computing time step is smaller than the plotting/printing interval, TSTEP.

The parameter UIC means use initial conditions. When UIC appears in the .TRAN control line, SPICE does not compute the quiescent operating point of the circuit prior to doing the transient analysis. SPICE will do one of two things:

1. If a .IC control line is not present, it will use the initial conditions specified on element lines (capacitors, inductors and semiconductors) as the starting point for its analysis.

2. If a .IC control line <u>is</u> present, it will use the node voltages on the .IC line to determine the initial conditions for elements.

EXAMPLES:

.TRAN 1U 80U
   A transient analysis will occur every 1 us from time zero to 80 us. 81 points will be plotted or printed.

.TRAN 2N 1000N
   The intent of this control line is to do a transient analysis every 2 nanoseconds, up to 1000 nanoseconds. However, an error will occur here because the number of points to be plotted or printed is 501, in excess of the SPICE default limit of 201. This problem is easily corrected by using the "LIMPTS =" command in a .OPTIONS control line. For example, the control line .OPTIONS LIMPTS = 600 would prevent an error message. See Appendix A.

.TRAN 5M 500M 100M 2M UIC
   Every 5 ms from time zero up to 500 ms a transient analysis will be done; however, results will not be stored (or plotted or printed) until 100 ms. The computing time step will be 2 ms (TSTEP is 5 ms, [TSTOP - TSTART]/50 = [500 ms - 100 ms]/50 = 8 ms). If 2M had not appeared in the .TRAN control line, the computing time step would have been the smaller of 5 ms and 8 ms, or 5 ms. Initial conditions will be used for the first transient analysis.

6.5   OPERATING POINT ANALYSIS
USING .OP CONTROL LINE

Control line format:

   .OP

That's it, plain and simple. There are no optional parameters one can specify. This control line instructs SPICE to solve for the operating point of the circuit with capacitors opened and inductors shorted, and to print detailed results of the operating point analysis. Such things as total cir-

cuit power dissipation and currents of voltage sources, as
well as small-signal parameters of semiconductors and nonlinear devices, are printed.

SPICE performs a DC operating point analysis without the
.OP control line when an AC analysis or transient analysis is
performed, but it does not print such detailed results.

6.6   TRANSFER FUNCTION ANALYSIS
USING .TF CONTROL LINE

Control line format:

.TF OUTPUTVAR INPUTSRC

OUTPUTVAR is a small-signal output variable (voltage or current); INPUTSRC is a small-signal input source (voltage or
current).   The .TF control line does a small-signal DC transfer function analysis, and will print the input resistance at
INPUTSRC, the output resistance at OUTPUTVAR, and some type
of transfer function (voltage gain, current gain, transresistance or transconductance) in the output file.

EXAMPLES:

.TF V(5) VIN
The specified output is the node 5 voltage, and the specified input is a voltage source already defined in the input
file as VIN. The transfer function would be a voltage gain.

.TF I(VIDRAIN) VGATE
The output is the current through voltage source VIDRAIN
(likely to be a dummy voltage source used as an ammeter in
the drain branch of a FET), while the input is a voltage
source already defined in the input file as VGATE. The transfer function would be current/voltage, or transconductance.

.TF V(12,15) IBASE
The output is the difference in voltage between nodes 12
and 15 (perhaps collectors in a BJT differential amplifier),
and the input is a current source previously defined in the
input file as IBASE. The transfer function is the quotient
of voltage/current, or transresistance.

6.7   SENSITIVITY ANALYSIS USING .SENS CONTROL LINE

Control line format:

    .SENS OV1 <OV2> ...

OV1 is the output variable such as node voltage or current through a voltage source. One output variable must be specified; others are optional. The DC small-signal sensitivity of each output variable to each circuit parameter will be determined. It is easy to fell entire forests due to massive amounts of paper consumed when this control line is used in any but the simplest circuits. A SPICE analysis of a simple two-transistor differential amplifier with 3 resistors, 2 BJTs and two batteries printed 34 lines of sensitivity information with only one output variable specified. Each BJT contributed 14 lines to the list.

EXAMPLES:

    .SENS V(5) I(VSUPPLY)

The small-signal DC sensitivity of the voltage at node 5 will be determined for each parameter in the circuit. In addition, the small-signal DC sensitivity of the current in voltage source VSUPPLY (defined elsewhere in the input file) will be determined for each parameter in the circuit.

6.8   DISTORTION ANALYSIS USING .DISTO CONTROL LINE

Control line format:

    .DISTO RLOAD INTER <SKW2 <REFPWR <SPW2>>>>

This control line must be used with a .AC control line. SPICE will determine the small-signal distortion characteristic of the circuit along with the AC small-signal sinusoidal steady-state analysis. RLOAD is the element name of the output resistor into which all the distortion power will be calculated. INTER is the interval at which a summary printout of the contributions of all non-linear devices to the total distortion is printed; if INTER is omitted or set to zero, no summary printout is done. For example, if INTER is set to 2, then a summary printout will occur at every other AC analysis frequency. Results of the distortion analysis can also be plotted and/or printed. See Chap. 7 for information on doing this.

The distortion analysis is done with one or two frequencies at the input. Let f1 be the frequency of the AC analysis; f2 is then equal to SKW2(f1). If SKW2 is omitted it defaults to 0.9, so that f2 = 0.9(f1). REFPWR is the reference power level used in calculating the distortion products. If omitted, REFPWR defaults to 1 mW (0 dBm). SPW2 is the amplitude of f2; if omitted, SPW2 defaults to 1.0.

The output resulting from a .DISTO control line will contain the following:

HD2, the amplitude of the second harmonic of f1, assuming that f2 is not present.

HD3, the amplitude of the third harmonic of f1, assuming that f2 is not present.

SIM2, the amplitude of the (f1 + f2) frequency component.

DIM2, the amplitude of the (f1 - f2) frequency component.

DIM3, the amplitude of the (2(f1) - f2) frequency component.

EXAMPLES:

.DISTO ROUTPUT 3 0.5
A distortion analysis will occur at every third frequency of the AC analysis. The second frequency, f2, will be 0.5(f1). The reference power level is 1 mW.

.DISTO RDRAIN 1
The distortion analysis will take place at each frequency of the AC analysis. The second frequency, f2, will be 0.9(f1) since no SKW2 value was specified. The reference power level is 1 mW by default.

6.9   NOISE ANALYSIS USING .NOISE CONTROL LINE

Control line format:

.NOISE OUTPUTV INPUTSRC NUMSUM

This control line must be used with a .AC control line. SPICE will perform a noise analysis of the circuit along with the AC small-signal steady-state analysis. OUTPUTV is a voltage which will be considered the output summing point for noise. INPUTSRC is the element name of an independent voltage or current source which will be the noise input reference. NUMSUM is the summary interval; it is the interval at which a summary printout of the contributions of all noise generators (resistors, semiconductors) is printed. If NUMSUM is omitted or set to zero, no summary printout is done. For example, if NUMSUM is set to 5, then a summary printout will occur at every fifth AC analysis frequency.

It is easy to consume large amounts of paper with this control line if NUMSUM is set to 1. Results of the noise analysis can also be plotted and/or printed. See Chap. 7 for information on doing this.

EXAMPLES:

.NOISE V(2,3) VIN1 10
A noise analysis will be done along with the AC analysis. The summary printout will list, at every tenth frequency, the output noise measured as the difference between nodes 2 and 3, and the equivalent input noise referenced to voltage source VIN1.

.NOISE V(17) IBIAS
A noise analysis is done with the output noise measured at node 17, and the equivalent input noise referenced to current source IBIAS. Although no summary printouts will occur, the output noise and equivalent input noise can be plotted and/or printed at each frequency of the AC analysis.

6.10 FOURIER ANALYSIS USING .FOUR CONTROL LINE

Control line format:

.FOUR FREQ OV1 <OV2 OV3 . . .>

A Fourier analysis can only be done in conjunction with a transient analysis (a .TRAN control line must appear in the same input file). The .FOUR control line instructs SPICE to determine the amplitudes of the DC component and the first nine frequency components of a waveform (the fundamental and second through eighth harmonics). Results are written to the output file without the need for .PLOT or .PRINT control lines.

FREQ is the fundamental frequency (determined by you). If the input to a circuit was a triangle waveform with a period of 5 ms, the fundamental frequency would be 1/(5 ms), or 200 Hz. OV1 . . . are the output variables on which a Fourier analysis is to be performed. It is important to note that the Fourier analysis is not performed over the entire time of the transient analysis. It is done only at the very end of the time, from (TSTOP-period) to TSTOP. Thus one must be sure to make the transient analysis at least 1/FREQ long. For example, for the .FOUR analysis to be meaningful, a circuit with a a 60 Hz sinusoid input should be at least 1/(60 Hz) or 16.7 ms long.

In order to achieve good accuracy in the Fourier analysis results, the maximum computing time step (TMAX in the .TRAN control line) should be set to period/100 or less.

EXAMPLES:

.FOUR 60 V(6) I(VSOURCE)
A Fourier analysis will be done on the voltage at node 6 and also on the current through VSOURCE. The user-specified fundamental frequency is 60 Hz.

.FOUR 2MEG V(8,2)
Using a fundamental frequency of 2 MHz, a Fourier analysis will be done on the voltage between nodes 8 and 2.

6.11 ANALYSIS AT DIFFERENT TEMPERATURES USING .TEMP CONTROL LINE

Control line format:

.TEMP T1 <T2 <T3 . . .>>

The .TEMP line sets the temperature(s) at which SPICE will simulate the circuit. T1, T2, . . . are the temperatures expressed in centigrade. If more temperatures than T1 are specified, all analyses are performed at each temperature. Temperatures less than -223.0 C will be ignored by SPICE.

SPICE will assume that the nominal circuit temperature, TNOM, is 27 C unless the TNOM option in the .OPTIONS control line is used to set another value. Model parameters for semiconductors are assumed to be valid at TNOM; TNOM also affects the resistance variation of temperature-dependent resistors.

EXAMPLES:

.TEMP 0 25 60 100
All analyses specified in the circuit input file will be performed at 0, 25, 60 and 100 degrees centigrade.

.TEMP -40
All analyses specified in the circuit input file will be performed at -40 degrees centigrade.

CHAPTER SUMMARY

SPICE analyses include DC, AC, transient, operating point, transfer function, sensitivity, distortion, noise and Fourier. In addition, the circuit temperature can be set to any value(s) with the .TEMP control line.

AC analysis can be done at many frequencies which are linearly- or logarithmically-spaced.

Some of the analyses can create very large output files; look at the output file on the monitor before printing it to save paper.

PROBLEMS

1. Refer to the circuit schematic diagram in Fig. 6.1, which has nodes notated for SPICE analysis.
  a. Write a SPICE input file describing the circuit, including a title line and a .END line.
  b. Analyze the circuit for the DC voltage across the 4K ohm resistor (including polarity), using conventional circuit analysis methods.
  c. Run a SPICE analysis using the input file from Prob. 1.a. How does your answer in 1.b compare with the results obtained from the SPICE analysis?

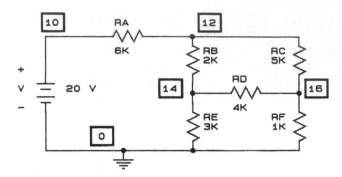

Figure 6.1 Circuit for Prob. 1.

2.   The   circuit   of   Fig.   6.2   contains   a   voltage-controlled
voltage source, E. The voltage across E, V(4,0), is given by

$$V(4,0) = 7*VA$$

where VA is the voltage across the 1 ohm resistor, V(2,1).
   a. Write a SPICE input file describing the circuit.
   b. Use SPICE to solve for the current through the 2 ohm
resistor (a dead voltage source is included in the circuit to serve as an ammeter).
   c. Modify the input file to determine the sensitivity of
the current through the 2 ohm resistor to the magnitude
of the 3 V battery. Run the analysis.

3.   A filter circuit used in frequency modulators is shown in
Fig. 6.3.   It is called a pre-emphasis circuit, and has an in-

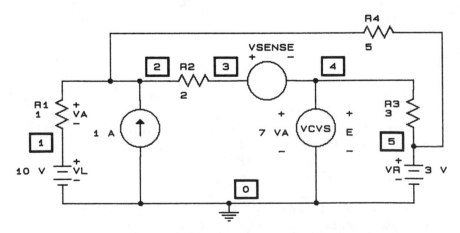

Figure 6.2 Circuit for Prob. 2.

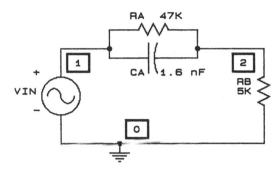

Figure 6.3   Circuit for Prob. 3.

teresting response as a function of frequency.   The response
is flat up to about 2 KHz, then rises at 20 dB/decade until
about 20 KHz, above which the response is flat again.   This
filter increases the amplitude of the frequencies above 2 KHz
prior to modulation.   In the receiver, after demodulation a
similar filter with the opposite response attenuates frequencies (and noise) above 2 KHz, improving the signal to noise
ratio at the output.

   a. Use SPICE to do an AC analysis from 200 Hz to 200 KHz,
   with logarithmically-spaced frequencies.   Plot the output
   voltage magnitude at node 2 versus frequency.

4.   Figure 6.4 is alleged to be a notch filter.   It consists
of   a   2-pole   high-pass   filter   in   cascade   with   a   2-pole   low-

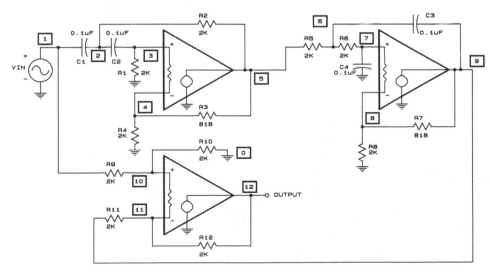

Figure 6.4   Circuit for Prob. 4.

pass filter, connected to the inverting input of a true differential amplifier.   The input signal feeds both the high-
pass filter and the non-inverting input of the differential
amplifier.   Each op-amp is modeled as an input resistance of
1E6 ohms, and a VCVS with a gain of 1E5 V/V.

    a. Write a SPICE input file describing the circuit.
Include an AC analysis which sweeps the frequency of VIN,
and a plot of the gain versus frequency.   For conven-
ience, make VIN have a magnitude of 1 V.   Choose a minimum and maximum frequency, as well as a type of sweep
(LIN, DEC and OCT are available) which will show the
notch frequency in a graph of output (node 12) voltage
versus frequency.

    b. Based on the result of 4.a, narrow the sweep range so
that the shape and depth of the notch are shown clearly.
What is the gain of the circuit at the center of the
notch?

5.   An array of 7 capacitors is shown in Fig. 6.5.   The
effective capacitance between nodes 1 and 0 is the unknown.

    a. Using the test circuit with an AC source and a 1 mH
inductor shown, sweep the source over an appropriate
range to make series resonance occur.   Look for resonance
by observing a plot of the voltage at node 1 (magnitude
and phase) versus frequency.   HINT: SPICE requires that
all nodes have a DC path to ground.   When nodes do not,
you have to put in a dummy resistor (1E9 ohm will do) in
order to run an analysis.

    b. As in Prob. 4.b, narrow the sweep range in the region
of resonance to get an accurate measure of the resonant
frequency.   Calculate effective capacitance using this
result.

6.   A 2-pole high-pass filter is shown in Fig. 6.6.   Its
response is very under-damped, due to excessive gain.   In
fact, just a bit more gain and it will be a free-running
oscillator.

    a. Model the op-amp as a 1E6 ohm input resistance and a
VCVS with a gain of 5E4 V/V.   Make the input voltage be a
pulse that goes from 0 V to 1 V at 0.5 ms, has a pulse
width of 0.4 ms, and rise and fall times of 1E-6 s.
Write a SPICE input file that will do a transient

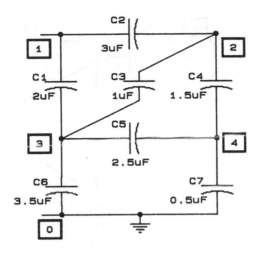

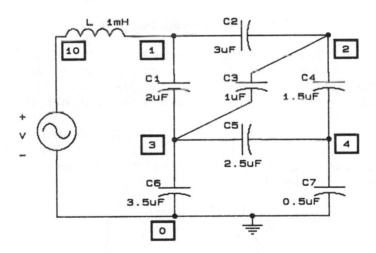

Figure 6.5 Circuits for Prob. 5.

analysis with time steps of 0.1 ms up to 5 ms.   Plot the
output voltage (node 5) and the input voltage (node 1)
versus time.   The damped oscillation of the filter at its
natural frequency should be very apparent.

7.   The Fourier analysis capability of SPICE can be used to
generate harmonic information for any periodic waveform.

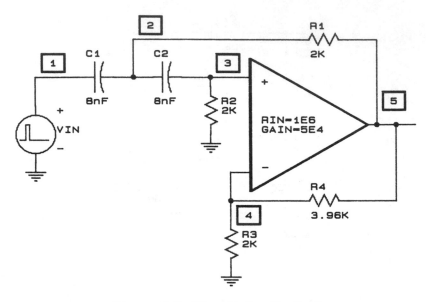

Figure 6.6 Circuit for Prob. 6.

Figure 6.7 shows a 1 KHz square wave voltage source, V, with a 1 ohm resistor to ground to satisfy the requirement that each node must be connected to at least two elements. The squarewave goes between -1 V and 1 V, and has rise and fall times of 1 ns.

　　a. Write an input file using a PULSE function for V. Do a transient analysis for 2 ms, with time steps of 1E-5 s. Use the .FOUR control line to produce the Fourier components at node 1.

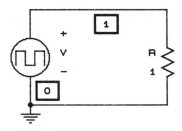

Figure 6.7 Circuit for Prob. 7.

Chapter 7

FORMATTING SPICE OUTPUT: PRINTING TABLES AND GRAPHS

7.1 INTRODUCTION

As we saw in Chap. 6 on the types of analyses possible with SPICE, some control lines will cause an automatic printing of analysis results. Among these control lines are: .OP (operating point), .TF (transfer function) and .SENS (sensitivity). Other kinds of analysis, such as .DC (non-linear DC), .AC (small-signal AC), .TRAN (non-linear transient analysis), .DISTO (distortion) and .NOISE (noise) do require that an additional control line be used to send results to the output file.

This chapter will cover two control lines, .PRINT and .PLOT, which, when additional parameters are included, will cause the analysis results to be written to the SPICE output file. .PRINT presents analysis results in tabular form, while .PLOT creates rectangular graphs of results. By investing a modest amount of time learning how these two control lines work, you will be able to make the output appear the way you want it to. Much time can be saved by letting SPICE do some of the major number crunching and then printing the final result. For example, SPICE can print and plot magnitudes in decibel form, as well as expressing complex numbers in both polar and rectangular forms.

7.2   USE OF THE .PRINT CONTROL LINE

7.2.1   .PRINT Format

The control line format is

.PRINT ANALTYPE OV1 <OV2 ... OV8>

ANALTYPE is one of five types of analysis

    DC
    AC
    TRAN
    DISTO
    NOISE

Notice that the ANALTYPE does not have a period (.) preceding it.   Up to 8 output variables can be listed together on one print control line, although only 5 output variables will fit on an 80 column-wide printer.   There is no limit to the number of print control lines in an input file.

OV stands for output variable. The format for OV1 through OV8 depends on the type of analysis being done. The .PRINT line will make a table of results appear in the output file. The first column will be the DC voltage (for DC analysis), the frequency (for AC analysis) or the time (for transient analysis).

7.2.2   DC and Transient Analysis
Output Variable Formats

Output variables for DC and transient analysis can be voltages at nodes compared to ground (node 0), voltages between two nodes, or currents flowing through independent voltage sources.   If an output variable has only one node listed, such as V(8), SPICE interprets that to be compared to ground.

EXAMPLES:

.PRINT DC V(1) V(5,7) I(VGATE)
The results of the DC analysis will be the magnitude of node 1 voltage (compared to ground), the voltage between nodes 5 and 7, and the current flowing through voltage source

VGATE. The output data table will have four columns, one for the first DC source listed in the .DC control line and three for the three output variables.

.PRINT TRAN V(6) I(VSENSING) V(4,2)

The results of the transient analysis will be the magnitude of node 6 voltage, the voltage between nodes 4 and 2, and the current flowing through voltage source VSENSING. The output data table will have four columns, one for the time steps specified in the .TRAN control line and three for the three output variables.

Remember that SPICE uses dead voltage sources as ammeters, so an appropriate name for such a current-sensing dead voltage source is VSENSING.

7.2.3   AC Analysis Output Variable Formats

Output variables for AC analysis can be voltages at nodes compared to ground (node 0), voltages between two nodes, or currents flowing through independent voltage sources.

For AC analyses, the output variables in the .PRINT control line could be the same as in both examples above for DC and transient analysis. If this were done, SPICE would print only the magnitude of each voltage or current. Since sinusoidal voltages and currents can be complex, SPICE allows for printing complex voltages as follows

V	magnitude
VR	real part
VI	imaginary part
VM	magnitude; V gives the same result.
VP	phase
VDB	20*LOG(magnitude)

Similarly, complex current can be specified as follows

I(VXXXXXX)	magnitude
IR(VXXXXXX)	real part
II(VXXXXXX)	imaginary part
IM(VXXXXXX)	magnitude
IP(VXXXXXX)	phase
IDB(VXXXXXX)	20*LOG(magnitude)

EXAMPLES:

.PRINT AC VM(3) VP(3) VR(3) VI(3) VDB(3)
This print control line will make SPICE create a table of the results of an AC analysis. The leftmost column will be all the frequencies specified by the .AC control line. The five columns to the right of the frequency column will contain the magnitude and phase of the voltage at node 3 (compared to node 0, or ground), the real and imaginary parts of the voltage at node 3, and the magnitude of the node 3 voltage in decibels. Thus, SPICE has presented the phasor voltage at node 3 in polar form, rectangular form, and in dB.

.PRINT AC IM(VSENSING) IP(VSENSING) IDB(VSHORT)
The magnitude and phase of the phasor current through voltage source VSENSING will be printed along with the magnitude of the current through voltage source VSHORT, expressed in decibels.

7.2.4   Noise and Distortion Analysis Output Variable Formats

Output variables for both noise and distortion analysis are structured differently from DC, AC and transient analysis. The kinds of output variables are

ONOISE	output noise
INOISE	equivalent input noise
HD2	2*f1 component (see .DISTO analysis, Chap. 6)
HD3	3*f1 component (ditto)
SIM2	f1 + f2 component (ditto)
DIM2	f1 - f2 component (ditto)
DIM3	2*f1 - f2 component (ditto)

and, associated with an output variable, any of the following may appear in parentheses:

R	real part
I	imaginary part
M	magnitude
P	phase
DB	20*LOG(magnitude)

If an output variable appears without R, I, M, P or DB in parentheses then the magnitude of that output variable will be printed.

EXAMPLES:

.PRINT NOISE INOISE ONOISE(DB)
Based on the noise analysis, the magnitude of the equivalent input noise and the output noise, in dB, will be printed at each frequency specified by the .NOISE and .AC control lines.

.PRINT DISTO HD2 HD3(DB) SIM2 DIM2
Based on the distortion analysis, the magnitude of the second harmonic, and third harmonic (in dB) distortion, as well as the (f1 + f2) and (f1 - f2) distortion components will be printed at each frequency specified by the .DISTO and .AC control lines.

7.3   USE OF THE .PLOT CONTROL LINE

In many instances during the circuit design process, exact values of circuit parameters are not needed. Very useful information may be obtained by looking at a graph of the parameters versus voltage (DC analysis), time (transient analysis) or frequency (AC analysis). Attempting to determine qualitative information, such as ripple in a Bode plot of filter gain, by examining tabular data can be time-consuming and ineffective. The .PLOT control line instructs SPICE to graph output data.

The plotting control line (.PLOT) follows the identical format as the print control line (.PRINT) except that the low and high plot limits can be specified for one or a group of output variables. If the optional plot limits are not specified, SPICE will automatically determine the maximum and minimum values of each variable being plotted, and will scale the output variable axis accordingly. The axis automatically may be made linear or logarithmic by SPICE. Although the automatic scaling results are not always to one's liking, it is a good idea to not specify plot limits the first time an analysis is run. Then, one can examine the plot in the output file, choose appropriate plot limits, and re-run the analysis with a revised input file.

SPICE uses a different symbol for each curve plotted. When two or more curves intersect, the plot symbol is a letter X at that point. The first output variable listed in a .PLOT control line will have its values printed in addition to being plotted. If you want the value of more than one output variable to be printed, a separate .PRINT control line can be used.

7.3.1   .PLOT Format

The control line format is

    .PLOT ANALTYPE OV1 <(PLO1,PHI1)> <OV2 <(PLO2,PHI2)>
    + ...   OV8 <(PLO8,PHI8)>>

Notice that the ANALTYPE does not have a period (.) preceding it. Up to 8 output variables can be listed together on one plot control line. To avoid very confusing graphs that are difficult to interpret, it is best to restrict the number of variables on one .PLOT line to two or three. There is no limit to the number of .PLOT control lines in an input file.

OV stands for output variable; the format for OV1 through OV8 depends on the type of analysis being done. The .PLOT line will make a graph of results appear in the output file; the abscissa, or horizontal axis, will be the DC voltage (for DC analysis), the frequency (for AC analysis) or the time (for transient analysis).

(PLO, PHI) are plot limits which are optional and can be put (with parentheses) after any of the output variables. If plot limits are used, they cause all output variables to their left (up to the previous output variable whose plot limits are specified) to be plotted using PLO as the lowest axis value and PHI as the highest axis value.

If the plotted values have sufficiently different ranges, SPICE will use and label more than one ordinate scale unless optional plot limits override this.

7.3.2   DC and Transient Analysis
Output Variable Formats

Output variables for DC and transient analysis can be voltages at nodes compared to ground (node 0), voltages between two nodes, or currents flowing through independent voltage sources.

EXAMPLE:

.PLOT DC V(1) (-5,12) V(5,7) I(VGATE)
The plotted results of the DC analysis will be the magnitude of node 1 voltage, the voltage between nodes 5 and 7,
and the current flowing through voltage source VGATE. The
output graph will have three plots; the abscissa (horizontal
axis) for all three will be the first DC source listed in the
.DC control line. The plots will be the voltage at node 1,
with scale limits of -5 V to +12 V, the voltage between nodes
5 and 7, and the current through voltage source VGATE. The
last two plots will be scaled automatically by SPICE.

EXAMPLE:

.PLOT TRAN V(6) I(VSENSING) V(4,2) (0,8)
The plotted results of the transient analysis will be the
magnitude of node 6 voltage, the current flowing through volt-
age source VSENSING, and the voltage between nodes 4 and 2.
The horizontal axis for the plots of the three output varia-
bles will be the time as specified in the .TRAN control line.
All three plots will have vertical scale limits of 0 and 8;
V(6) and V(4,2) will be volts, ISENSING will be amps.

7.3.3   AC Analysis Output Variable Formats

Output variables for AC analysis can be voltages at nodes compared to ground (node 0), voltages between two nodes, or currents flowing through independent voltage sources.

For AC analyses, the output variables in the .PLOT control
line could be the same as in both examples above for DC and
transient analysis. If this were done, SPICE would plot only
the magnitude of each voltage or current. Since sinusoidal
voltages and currents can be complex, SPICE allows for plotting complex voltages as follows:

V   magnitude
VR   real part
VI   imaginary part
VM   magnitude; V gives the same result.
VP   phase
VDB   20*LOG(magnitude)

Similarly, complex current can be specified as follows:

I(VXXXXXX)	magnitude
IR(VXXXXXX)	real part
II(VXXXXXX)	imaginary part
IM(VXXXXXX)	magnitude
IP(VXXXXXX)	phase
IDB(VXXXXXX)	20*LOG(magnitude)

If the AC input voltage to a circuit has a magnitude of 1 V, use of VDB for an output voltage will produce a Bode plot, providing that a logarithmic sweep of frequency is done.

EXAMPLE:

.PLOT AC VM(3) VP(3)
This plot control line will make SPICE create a graph of the results of an AC analysis. The abscissa will be frequency, as specified by the .AC control line. The two plots will be the magnitude and phase of the voltage at node 3 (compared to node 0, or ground). The two vertical axes of the graph will be scaled automatically.

EXAMPLE:

.PLOT AC IM(VSENSING) (0,10M) VDB(35,19)
The magnitude of the phasor current through voltage source VSENSING will be plotted along with the magnitude of the voltage between nodes 35 and 19, expressed in decibels. The scale of the current plot will be 0 to 1E-02, while the voltage plot will be scaled automatically.

7.3.4   Noise and Distortion Analysis Output Variable Formats

Output variables for both noise and distortion analysis are structured differently from DC, AC and transient analysis. The kinds of output variables are:

ONOISE	output noise
INOISE	equivalent input noise
HD2	2*f1 component (see .DISTO analysis, Chap. 6)
HD3	3*f1 component (ditto)

SIM2	f1 + f2 component (ditto)
DIM2	f1 - f2 component (ditto)
DIM3	2*f1 - f2 component (ditto)

and associated with an output variable any of the following may appear in parentheses:

R	real part
I	imaginary part
M	magnitude
P	phase
DB	20*LOG(magnitude)

If an output variable appears without R, I, M, P or DB in parentheses then the magnitude of that output variable will be plotted.

EXAMPLE:

.PLOT NOISE INOISE ONOISE(DB)
Based on the noise analysis, the magnitude of the equivalent input noise and the output noise, in dB, will be plotted versus frequency as specified by the .NOISE and .AC control lines.

EXAMPLE:

.PLOT DISTO HD2 SIM2 (0,0.1) DIM2 HD3(DB)
Based on the distortion analysis, the magnitude of the second harmonic distortion, in addition to the (f1 + f2) and (f1 - f2) distortion components and third harmonic distortion (in dB), will be plotted versus frequency as specified by the .DISTO and .AC control lines. The scale for the first two plots will be from 0 to 0.1, the last two will be scaled automatically by SPICE.

7.4   INTERPRETING SPICE OUTPUT INFORMATION

7.4.1   Interpreting .PRINT Numerical Data

The .PRINT control line creates tables of numerical information. Results are printed in scientific notation. For example, the number zero would appear in a table as .000E+00, where E indicates exponent of 10. So 7.345E+01 means 73.45, and 1.286E-02 means 0.01286.   Be careful when reading results to avoid errors that can be off by one or more orders of magnitude.   It is not hard to mistakenly interpret a current of 5.400E-7 A as 54 uA, when in fact it is 0.54 uA.

The example in Section 7.4.3 will show the how to use the .PRINT control line to create a table of results.

7.4.2   Interpreting .PLOT Numerical Data

The .PLOT control line creates a graph of numerical information.   The first variable listed in a .PLOT line also has its value printed in scientific notation, with D meaning exponent of 10.   It would be redundant to have the following two control lines

```
.PRINT      AC  VM(8,3)
.PLOT       AC  VM(8,3)
```

The .PRINT line will create a table with two columns: frequency (as specified in the .AC control line) and magnitude of the voltage between nodes 8 and 3.   The .PLOT line will create those same two columns as well as plotting VM(8,3) versus frequency.   Thus, the same information could be obtained with the .PLOT line alone.

When more than one variable is plotted, the .PLOT line will create only two columns.   In this situation a .PRINT line could be used to print values of all data needed.

7.4.3   Example of .PRINT and .PLOT

The circuit in Fig. 7.1 shows a triangle voltage source connected to a resistor in series with an inductor.   The triangle waveform is made by using an independent voltage source PULSE function, with the pulse width essentially zero.   Figure 7.2 shows the input file LR.CIR, with the rise time set to 0.5 ms, fall time of 0.5 ms, and pulse width of 1 ns (since SPICE gives erroneous results if zero is used).   Chapter 12.4 shows how to make a triangle waveform with the piecewise linear (PWL) source.

The control lines in LR.CIR instruct SPICE to print and plot the results of the transient analysis over two periods of the input voltage (a period is 1 ms).   The .PRINT line causes a table to be made of all possible voltages.   The

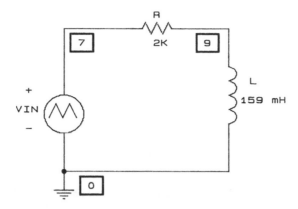

Figure 7.1 Series R-C Circuit with Triangle Input Voltage.

first .PLOT control line has plot limits of 0 and 1 V for V(7). The second .PLOT line does not specify plot limits; SPICE will determine the scale automatically for V(9). The third .PLOT line will graph both parameters, and SPICE will determine the scale automatically for each plot.

Figure 7.3 is the output file LR.OUT, which contains a table and three graphs. The table has four columns: time, and the three voltages specified. The first graph shows the triangle voltage, V(7), starting at 0 V at 0 ms, rising to 1 V at 500 us and falling back to 0 V at 1 ms. The triangle is repeated from 1 ms to 2 ms. Notice that the magnitude of V(7) is printed next to each time value along the time axis of the graph.

The second graph shows the voltage across the inductor, V(9). SPICE has scaled the voltage axis with good results. The third graph shows both V(9) and V(7), for which the plot symbols are asterisks (*) and plus signs (+) respectively.

```
LR.CIR  TRIANGLE WAVEFORM MADE WITH PULSE FUNCTION, L-R HP FILTER
VIN     7 0  PULSE(0  1  0  .5M  .5M  1N 1M)
*       THE LINE ABOVE IS A 1 VPP TRIANGLE AT 1KHZ
R       7 9 2K
L       9 0 159M
.TRAN 50U 2M
.PRINT  TRAN  V(7)  V(9)  V(7,9)
.PLOT TRAN V(7)   (0,1)
.PLOT TRAN V(9)
.PLOT TRAN V(9) V(7)
.OPTIONS NOPAGE
.END
```

Figure 7.2 Input File LR.CIR.

```
********11/22/87******** Demo PSpice (May 1986) *******21:27:49********


LR.CIR  TRIANGLE WAVEFORM MADE WITH PULSE FUNCTION, L-R HP FILTER


****    CIRCUIT DESCRIPTION


**********************************************************************


VIN    7  Ø  PULSE(Ø  1  Ø  .5M  .5M  1N 1M)
*      THE LINE ABOVE IS A 1 VPP TRIANGLE AT 1KHZ
R      7  9  2K
L      9  Ø  159M
.TRAN  50U  2M
.PRINT TRAN  V(7)  V(9)  V(7,9)
.PLOT TRAN V(7)   (Ø,1)
.PLOT TRAN V(9)
.PLOT TRAN V(9) V(7)
.OPTIONS NOPAGE
.END

****    INITIAL TRANSIENT SOLUTION      TEMPERATURE =   27.ØØØ DEG C

 NODE   VOLTAGE      NODE   VOLTAGE


( 7)     .ØØØØ    ( 9)     .ØØØØ
```

Figure 7.3 Output File LR.OUT.

```
****     TRANSIENT ANALYSIS                TEMPERATURE =   27.000 DEG C


     TIME       V(7)         V(9)        V(7,9)


    .000E+00    .000E+00     .000E+00     .000E+00
   5.000E-05   1.000E-01    7.257E-02    2.743E-02
   1.000E-04   2.000E-01    1.134E-01    8.663E-02
   1.500E-04   3.000E-01    1.356E-01    1.644E-01
   2.000E-04   4.000E-01    1.464E-01    2.536E-01
   2.500E-04   5.000E-01    1.523E-01    3.477E-01
   3.000E-04   6.000E-01    1.555E-01    4.445E-01
   3.500E-04   7.000E-01    1.572E-01    5.428E-01
   4.000E-04   8.000E-01    1.580E-01    6.420E-01
   4.500E-04   9.000E-01    1.585E-01    7.415E-01
   5.000E-04   1.000E+00    1.587E-01    8.413E-01
   5.500E-04   9.000E-01    1.299E-02    8.870E-01
   6.000E-04   8.000E-01    6.740E-02    8.674E-01
   6.500E-04   7.000E-01   -1.109E-01    8.110E-01
   7.000E-04   6.000E-01   -1.343E-01    7.343E-01
   7.500E-04   5.000E-01   -1.458E-01    6.459E-01
   8.000E-04   4.000E-01   -1.520E-01    5.520E-01
   8.500E-04   3.000E-01   -1.553E-01    4.553E-01
   9.000E-04   2.000E-01   -1.571E-01    3.571E-01
   9.500E-04   1.000E-01   -1.580E-01    2.580E-01
   1.000E-03   1.863E-06   -1.585E-01    1.585E-01
   1.050E-03   1.000E-01   -1.289E-02    1.129E-01
   1.100E-03   2.000E-01    7.020E-02    1.298E-01
   1.150E-03   3.000E-01    1.112E-01    1.888E-01
   1.200E-03   4.000E-01    1.336E-01    2.664E-01
   1.250E-03   5.000E-01    1.457E-01    3.543E-01
   1.300E-03   6.000E-01    1.522E-01    4.478E-01
   1.350E-03   7.000E-01    1.553E-01    5.447E-01
   1.400E-03   8.000E-01    1.571E-01    6.429E-01
   1.450E-03   9.000E-01    1.580E-01    7.420E-01
   1.500E-03   1.000E+00    1.585E-01    8.415E-01
   1.550E-03   9.000E-01    1.285E-02    8.872E-01
   1.600E-03   8.000E-01   -6.747E-02    8.675E-01
   1.650E-03   7.000E-01   -1.110E-01    8.110E-01
   1.700E-03   6.000E-01   -1.343E-01    7.343E-01
   1.750E-03   5.000E-01   -1.459E-01    6.459E-01
   1.800E-03   4.000E-01   -1.520E-01    5.520E-01
   1.850E-03   3.000E-01   -1.553E-01    4.553E-01
   1.900E-03   2.000E-01   -1.571E-01    3.571E-01
   1.950E-03   1.000E-01   -1.580E-01    2.580E-01
   2.000E-03   4.000E-06   -1.585E-01    1.585E-01
```

Figure 7.3 (Continued).

```
****    TRANSIENT ANALYSIS                  TEMPERATURE =   27.000 DEG C

   TIME      V(7)

                      .000D+00    2.500D-01   5.000D-01   7.500D-01   1.000D+00
                  - - - - - - - - - - - - - - - - - - - - - - - - - - - - - - -
  .000D+00   .000D+00 *          .           .           .           .
 5.000D-05  1.000D-01 .     *    .           .           .           .
 1.000D-04  2.000D-01 .          *  .        .           .           .
 1.500D-04  3.000D-01 .          .   *       .           .           .
 2.000D-04  4.000D-01 .          .        *  .           .           .
 2.500D-04  5.000D-01 .          .           *           .           .
 3.000D-04  6.000D-01 .          .           .       *   .           .
 3.500D-04  7.000D-01 .          .           .           *           .
 4.000D-04  8.000D-01 .          .           .           .    *      .
 4.500D-04  9.000D-01 .          .           .           .        *  .
 5.000D-04  1.000D+00 .          .           .           .           *
 5.500D-04  9.000D-01 .          .           .           .        *  .
 6.000D-04  8.000D-01 .          .           .           .    *      .
 6.500D-04  7.000D-01 .          .           .           *           .
 7.000D-04  6.000D-01 .          .           .       *   .           .
 7.500D-04  5.000D-01 .          .           *           .           .
 8.000D-04  4.000D-01 .          .        *  .           .           .
 8.500D-04  3.000D-01 .          .   *       .           .           .
 9.000D-04  2.000D-01 .          *  .        .           .           .
 9.500D-04  1.000D-01 .     *    .           .           .           .
 1.000D-03  1.863D-06 *          .           .           .           .
 1.050D-03  1.000D-01 .     *    .           .           .           .
 1.100D-03  2.000D-01 .          *  .        .           .           .
 1.150D-03  3.000D-01 .          .   *       .           .           .
 1.200D-03  4.000D-01 .          .        *  .           .           .
 1.250D-03  5.000D-01 .          .           *           .           .
 1.300D-03  6.000D-01 .          .           .       *   .           .
 1.350D-03  7.000D-01 .          .           .           *           .
 1.400D-03  8.000D-01 .          .           .           .    *      .
 1.450D-03  9.000D-01 .          .           .           .        *  .
 1.500D-03  1.000D+00 .          .           .           .           *
 1.550D-03  9.000D-01 .          .           .           .        *  .
 1.600D-03  8.000D-01 .          .           .           .    *      .
 1.650D-03  7.000D-01 .          .           .           *           .
 1.700D-03  6.000D-01 .          .           .       *   .           .
 1.750D-03  5.000D-01 .          .           *           .           .
 1.800D-03  4.000D-01 .          .        *  .           .           .
 1.850D-03  3.000D-01 .          .   *       .           .           .
 1.900D-03  2.000D-01 .          *  .        .           .           .
 1.950D-03  1.000D-01 .     *    .           .           .           .
 2.000D-03  4.000D-06 *          .           .           .           .
                  - - - - - - - - - - - - - - - - - - - - - - - - - - - - - - -
```

Figure 7.3 (Continued).

```
****    TRANSIENT ANALYSIS              TEMPERATURE =   27.000 DEG C


    TIME      V(9)

                 -2.000D-01   -1.000D-01    .000D+00   1.000D-01    2.000D-01
           - - - - - - - - - - - - - - - - - - - - - - - - - - - - - - - - -
   .000D+00    .000D+00 .             .             *             .             .
  5.000D-05   7.257D-02 .             .             .          *  .             .
  1.000D-04   1.134D-01 .             .             .             . *           .
  1.500D-04   1.356D-01 .             .             .             .    *        .
  2.000D-04   1.464D-01 .             .             .             .     *       .
  2.500D-04   1.523D-01 .             .             .             .      *      .
  3.000D-04   1.555D-01 .             .             .             .      *      .
  3.500D-04   1.572D-01 .             .             .             .      *      .
  4.000D-04   1.580D-01 .             .             .             .       *     .
  4.500D-04   1.585D-01 .             .             .             .       *     .
  5.000D-04   1.587D-01 .             .             .             .       *     .
  5.500D-04   1.299D-02 .             .             .       *     .             .
  6.000D-04  -6.740D-02 .             .         *   .             .             .
  6.500D-04  -1.109D-01 .             .  *.         .             .             .
  7.000D-04  -1.343D-01 .         *    .            .             .             .
  7.500D-04  -1.458D-01 .      *       .            .             .             .
  8.000D-04  -1.520D-01 .    *         .            .             .             .
  8.500D-04  -1.553D-01 .    *         .            .             .             .
  9.000D-04  -1.571D-01 .    *         .            .             .             .
  9.500D-04  -1.580D-01 .   *          .            .             .             .
  1.000D-03  -1.585D-01 .   *          .            .             .             .
  1.050D-03  -1.289D-02 .             .          *  .             .             .
  1.100D-03   7.020D-02 .             .             .        *    .             .
  1.150D-03   1.112D-01 .             .             .             .*            .
  1.200D-03   1.336D-01 .             .             .             .   *         .
  1.250D-03   1.457D-01 .             .             .             .    *        .
  1.300D-03   1.522D-01 .             .             .             .     *       .
  1.350D-03   1.553D-01 .             .             .             .     *       .
  1.400D-03   1.571D-01 .             .             .             .     *       .
  1.450D-03   1.580D-01 .             .             .             .      *      .
  1.500D-03   1.585D-01 .             .             .             .      *      .
  1.550D-03   1.285D-02 .             .             .       *     .             .
  1.600D-03  -6.747D-02 .             .         *   .             .             .
  1.650D-03  -1.110D-01 .             .  *.         .             .             .
  1.700D-03  -1.343D-01 .         *    .            .             .             .
  1.750D-03  -1.459D-01 .      *       .            .             .             .
  1.800D-03  -1.520D-01 .    *         .            .             .             .
  1.850D-03  -1.553D-01 .    *         .            .             .             .
  1.900D-03  -1.571D-01 .    *         .            .             .             .
  1.950D-03  -1.580D-01 .   *          .            .             .             .
  2.000D-03  -1.585D-01 .   *          .            .             .             .
           - - - - - - - - - - - - - - - - - - - - - - - - - - - - - - - - -
```

Figure 7.3 (Continued).

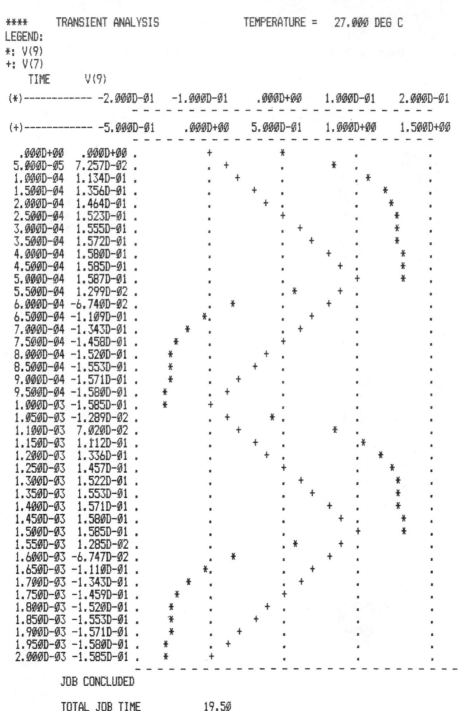

```
****    TRANSIENT ANALYSIS                TEMPERATURE =   27.000 DEG C
LEGEND:
*: V(9)
+: V(7)
   TIME        V(9)
(*)------------- -2.000D-01   -1.000D-01    .000D+00    1.000D-01    2.000D-01
                             - - - - - - - - - - - - - - - - - - - - -
(+)------------- -5.000D-01    .000D+00    5.000D-01    1.000D+00    1.500D+00
                             - - - - - - - - - - - - - - - - - - - - -
    .000D+00   .000D+00 .             +            *         .          .
  5.000D-05  7.257D-02 .           .   +         .         *  .          .
  1.000D-04  1.134D-01 .           .      +      .          . *          .
  1.500D-04  1.356D-01 .           .        +    .          .      *     .
  2.000D-04  1.464D-01 .           .          +  .          .       *    .
  2.500D-04  1.523D-01 .           .           + .          .        *   .
  3.000D-04  1.555D-01 .           .          . +           .        *   .
  3.500D-04  1.572D-01 .           .          .  +          .        *   .
  4.000D-04  1.580D-01 .           .          .      +      .         *  .
  4.500D-04  1.585D-01 .           .          .       +     .         *  .
  5.000D-04  1.587D-01 .           .          .        +    .         *  .
  5.500D-04  1.299D-02 .           .          . *          +          .  .
  6.000D-04 -6.740D-02 .           .   *      .          + .          .  .
  6.500D-04 -1.109D-01 .        *.           .          +  .          .  .
  7.000D-04 -1.343D-01 .      *      .         .      +    .          .  .
  7.500D-04 -1.458D-01 .    *        .          +         .          .  .
  8.000D-04 -1.520D-01 .   *         .      +    .          .          .  .
  8.500D-04 -1.553D-01 .   *         .     +     .          .          .  .
  9.000D-04 -1.571D-01 .   *         .   +       .          .          .  .
  9.500D-04 -1.580D-01 .  *         . +          .          .          .  .
  1.000D-03 -1.585D-01 .  *       +            .          .          .  .
  1.050D-03 -1.289D-02 .           . +        *.          .          .  .
  1.100D-03  7.020D-02 .           .  +        .          *          .  .
  1.150D-03  1.112D-01 .           .       +   .          . *        .  .
  1.200D-03  1.336D-01 .           .        +  .          .    *     .  .
  1.250D-03  1.457D-01 .           .          +.          .      *   .  .
  1.300D-03  1.522D-01 .           .          . +         .        * .  .
  1.350D-03  1.553D-01 .           .          .   +       .        * .  .
  1.400D-03  1.571D-01 .           .          .      +    .        *  .
  1.450D-03  1.580D-01 .           .          .       +   .         * .
  1.500D-03  1.585D-01 .           .          .        +  .         * .
  1.550D-03  1.285D-02 .           .          . *         +          .  .
  1.600D-03 -6.747D-02 .           .   *      .          + .          .  .
  1.650D-03 -1.110D-01 .        *.            .          +  .          .  .
  1.700D-03 -1.343D-01 .      *      .         .      +     .          .  .
  1.750D-03 -1.459D-01 .    *        .          +         .          .  .
  1.800D-03 -1.520D-01 .   *         .      +    .          .          .  .
  1.850D-03 -1.553D-01 .   *         .     +     .          .          .  .
  1.900D-03 -1.571D-01 .   *         .   +       .          .          .  .
  1.950D-03 -1.580D-01 .  *         . +          .          .          .  .
  2.000D-03 -1.585D-01 .  *       +            .          .          .  .
                             - - - - - - - - - - - - - - - - - - - - -

         JOB CONCLUDED

         TOTAL JOB TIME        19.50
```

Figure 7.3 (Continued).

Notice that only the magnitude of V(9) is printed along the time axis. Each variable has its own voltage axis scale. Also, the automatic voltage axis scaling of V(7) results in a compressed plot which is not as useful as the plot of V(7) in the first graph. This could be improved by changing the third .PLOT line in Fig. 7.2 to

.PLOT TRAN V(9) (-0.2, 0.2) V(7) (0,1)

Recall that plot limits affect all output variables to the left of the plot limit specification, unless different plot limits were specified. Thus, if (-0.2, 0.2) was not present, both V(9) and V(7) would use the (0,1) plot limit. This would chop off the bottom part of the V(9) waveform.

7.4.4   Obtaining Higher–Quality Graphs

As we can see from Fig. 7.3, graphs done on a line printer using standard symbols leave something to be desired. The graph in Fig. 7.4 certainly is easier to interpret and more

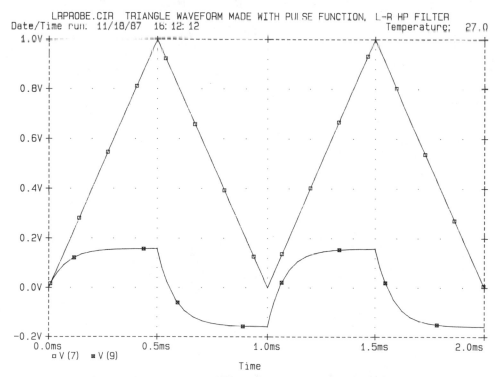

Figure 7.4 Graph of Source Voltage and Inductor Voltage Done with Graphics Post-Processor.

like publication-quality graphics; the standard SPICE program cannot directly produce this kind of graph. Figure 7.4 was made with a graphics post-processor program. Several suppliers of SPICE versions for personal computers offer graphics post-processors that produce fine-quality graphical output.

The particular graphics post-processor used was PROBE (by Microsim Corporation), which can be purchased for use with PSPICE. In addition to the professional-looking graphs that result, PROBE can perform mathematical operations on output variables and graph the results. For example, one could multiply the resistor voltage by itself, and divide the product by the value of the resistance. The resulting graph would be resistor power (V*V/R) versus time. This is shown in Fig. 7.5.

Some of the graphs produced by SPICE in this book are not easily interpreted as printed. For this reason, the plot points were connected by hand using drafting instruments to enhance the readability of the graphs.

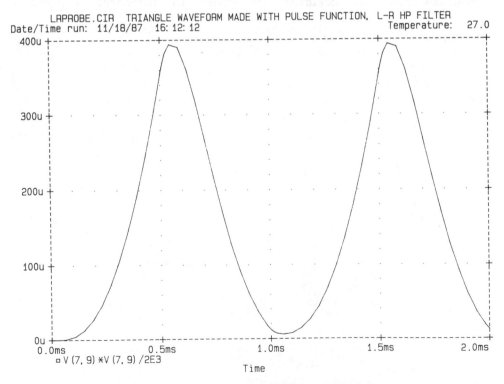

Figure 7.5 Graph of Resistor Power Done with Graphics Post-Processor.

CHAPTER SUMMARY

Results of DC, AC, transient, distortion and noise analyses can be printed in tabular format and/or plotted by SPICE.

Complex AC analysis results can be expressed in two ways: polar (VM and VP) and rectangular (VR and VI).   Also the decibel expression of a voltage or current magnitude can be computed by SPICE.

Graphs will be scaled automatically by SPICE.   The scaling can be user-specified by use of optional plot limits.

While the graphs produced by SPICE are very useful, graphics post-processors are available from several suppliers of SPICE versions for PCs which produce professional-looking results.

Chapter 8

SEMICONDUCTORS IN SPICE

8.1 INTRODUCTION

SPICE has built-in models for four semiconductor devices: diodes, bipolar junction transistors (BJTs), field-effect transistors (FETs) and metal-oxide semiconductor field-effect transistors (MOSFETs). These internal models are typical of such devices when created on an integrated circuit. Each type has a large number of parameters, and SPICE has specific default values for each parameter. For example, the SPICE model for a BJT has 40 parameters including forward beta (DC current gain), reverse beta (current gain when you interchange collector and emitter), emitter resistance, and a number of others little known to most people who practice electronics. The default value (the value SPICE assigns to the parameter <u>unless you specify otherwise</u>) for forward beta is 100, while reverse beta defaults to 1.

In this chapter the element lines for the four kinds of semiconductor devices known by SPICE will be presented, along with the .MODEL control line which must be used whenever a semiconductor element line is included in the input file. Before you become overwhelmed with the prospect of modeling a circuit with semiconductors, rest assured that most of the time you need not deal with the vast majority of the myriad parameters. The only data needed to specify each semiconductor are a single element line and a single, short .MODEL line which specifies only those few parameters which you feel are important. In fact, any number of element lines for the same kind of semiconductor device (e.g., an NPN BJT transistor) can "share" the same single .MODEL line.

8.2   DESCRIBING DIODES TO SPICE

8.2.1   Diode Element Line

The general form for an element line for a junction diode is

> DXXXXXX N+ N- MODNAME <AREA> <OFF> <IC=VD>

where DXXXXXX is the name of the diode, N+ is the node to which the anode is connected, N- is the node to which the cathode is connected and MODNAME is the model name which is used in an associated .MODEL control line. All of these are required in the diode element line. The optional parameters are:

> AREA - the area factor which determines how many of the diode model MODNAME are put in parallel to make one DXXXXXX. The area parameter will affect IS, RS, CJO and IBV in the model of the diode (see Section 8.2.3). If not specified, area defaults to 1.

> OFF - an initial condition of DXXXXXX for the DC analysis.

> IC=VD - causes SPICE to use VD as the initial condition for diode voltage instead of the quiescent operating point diode voltage when a transient analysis is done.

8.2.2   Diode Model Line

The general form for a .MODEL line for a junction diode is

> .MODEL MODNAME D<(PAR1=PVAL1 PAR2=PVAL2 ...)>

where MODNAME is the model name given to a diode in an element line, and D tells SPICE that the semiconductor being mod-

eled is a diode and not a BJT, FET or MOSFET. PAR is the parameter name, which can be any of those in the parameter list
in Section 8.2.3, and PVAL is the value of that parameter.
Note that the PAR and PVAL parts of the .MODEL line are optional; the minimum specification of a diode is simply the
letter D.

EXAMPLE:

    DRECTIFY 7 15 HIPOWER
    .MODEL HIPOWER D(RS = 0.5 BV = 400 IBV = 50M)
    The diode named DRECTIFY has its anode at node 7 and its
cathode at node 15. Its model name is HIPOWER. The model
named HIPOWER includes 0.5 ohm of ohmic resistance, a reverse
breakdown voltage of 400 V, and a current at reverse
breakdown of 50 mA.

EXAMPLE:

    DBIG10 9 4 DHEFTY
    DBIG20 5 9 DHEFTY 2.5
    .MODEL DHEFTY D(IS = 2E-14 BV = 8E2 IBV = 0.1)
    Two diodes are described in these three lines. The anode
of DBIG10 and the cathode of DBIG20 are connected to node 9.
Both use the single DHEFTY model line. DHEFTY has a saturation current of 0.02 pA, a reverse breakdown voltage of 800
volts, and a current at reverse breakdown of 100 mA. However, the diodes are not the same, even though both use the
DHEFTY model. Before you give up in confusion, notice the
DBIG20 element line which has the number 2.5 in it. 2.5 is
the area factor of DBIG20. This means that the physical area
of diode DBIG20 will be 2.5 times the area of the DHEFTY
diode, and the IS, RS, CJO, and IBV parameters of diode
DBIG20 will be different from the IS, RS, CJO and IBV
parameters of diode DBIG10.

EXAMPLE:

    DREF 1 6 TYPE17
    .MODEL TYPE17 D(BV = 6.2 IBV = 10M)
    The diode named DREF has its anode at node 1 and its cath-
ode at node 6. It appears to be a Zener diode, due to its
low reverse breakdown voltage of 6.2 V.

8.2.3   Diode Model Parameters

#	Name	Parameter	Units	Default	Example	Area
1	IS	saturation current	A	1.0E-14	2.0E-14	*
2	RS	ohmic resistance	Ohm	0	6	*
3	N	emission coefficient	-	1	0.9	
4	TT	transit-time	sec	0	0.3Ns	
5	CJO	zero-bias junction capacitance	F	0	1.5PF	*
6	VJ	junction potential	V	1	0.8	
7	M	grading coefficient	-	0.5	0.6	
8	EG	activation energy	eV	1.11	1.11 Si 0.69 Sbd 0.67 Ge	
9	XTI	saturation-current temp. exp	-	3.0	3.0 jn 2.0 Sbd	
10	KF	flicker noise coefficient	-	0		
11	AF	flicker noise exponent	-	1		
12	FC	coefficient for forward-bias depletion capacitance formula	-	0.5		
13	BV	reverse breakdown voltage	V	infinite	200	
14	IBV	current at breakdown voltage	A	1.0E-3		*

In the table above, "Sbd" stands for Schottky barrier diode, "Si" means silicon, "Ge" represents germanium, and "jn" indicates junction.

Notice that if no parameters are specified in the .MODEL MODNAME D model line, then ohmic resistance (RS), transit time (TT), flicker noise coefficient (KF) and zero-bias junction capacitance (CJO) all default to zero. Also, the default value of the reverse breakdown voltage (BV) is infinity. If BV or IBV are specified in the diode model line, positive values should be used.

8.3   DESCRIBING BIPOLAR JUNCTION TRANSISTORS TO SPICE

8.3.1   BJT Element Line

The general form for an element line for a bipolar junction transistor is

QXXXXXX NC NB NE <NS> MODNAME <AREA> <OFF> <IC=VBE,VCE>

where QXXXXXX is the name of the transistor. NC, NB and NE are the nodes to which the collector, base and emitter are connected, respectively.   MODNAME is the model name which is used in an associated .MODEL control line.   All of these are required in the BJT element line.   The optional parameters are

> NS - the node to which the substrate is connected; if omitted, NS defaults to node 0.

> AREA - the area factor which determines how many of the BJT model MODNAME are put in parallel to make one QXXXXXX.

> OFF - an initial condition of QXXXXXX for the DC analysis.

> IC=VBE,VCE - a specification of initial conditions, for use with the UIC option of the .TRAN control line.   This causes SPICE to use VBE and VCE as the initial conditions for base-emitter and collector-emitter voltages, instead of the quiescent operating point junction voltages when a transient analysis is done.

8.3.2   BJT Model Line

The general forms for model lines for bipolar junction transistors are

.MODEL MODNAME NPN<(PAR1=PVAL1 PAR2=PVAL2 ...)>
.MODEL MODNAME PNP<(PAR1=PVAL1 PAR2=PVAL2 ...)>

where MODNAME is the model name given to a BJT in an element line, and NPN or PNP tells SPICE that the semiconductor being

modeled is an NPN or PNP BJT and not a diode, FET or MOSFET.
PAR is the parameter name, which can be any of those in the
parameter list in Section 8.3.3, and PVAL is the value of
that parameter.   Note that the PAR and PVAL parts of the
.MODEL line are optional; the minimum specification of a BJT
is simply NPN or PNP.

There are 40 parameters that can be specified in the
.MODEL control line for a BJT. The model is adapted from the
integral charge control model of Gummel and Poon.   If certain
parameters are not specified, the model will become the simp-
ler Ebers-Moll model.

EXAMPLE:

QBUFFER 4 10 5 SMALLSIG
.MODEL SMALLSIG NPN(BF = 140)
Bipolar junction transistor QBUFFER has its collector at
node 4, its base at node 10 and its emitter at node 5.   The
substrate of QBUFFER, by default, is node 0.   This is an NPN
transistor with all the default parameters listed in Section
8.3.3, except that its forward beta is set to 140.

EXAMPLE:

QTOP   3 8 9 HIPOWER 8
QMID   1 2 4 HIPOWER
QBOTTOM   6 7 9 HIPOWER 8
.MODEL HIPOWER NPN(BF = 50)
Bipolar junction transistors QTOP and QBOTTOM have area
factors of 8, which means their physical areas will be 8
times the area of QMID.   All parameters in Section 8.3.3 with
an asterisk (*) in the area column will be affected by the
area factor.   By default, QMID has an area factor of 1.

All three NPN BJTs have a forward beta of 50.   Note that
only one .MODEL line is needed for these three transistors.
Any number of transistors that have the same MODNAME can use
the same .MODEL line.

EXAMPLE:

QINPUT 10 11 12 MOD15
.MODEL MOD15 PNP
BJT QINPUT has its collector, base and emitter connected
to nodes 10, 11 and 12, respectively.   It is a PNP type, with
all the default parameters listed in Section 8.3.3.

8.3.3   BJT Model Parameters

Modified Gummel-Poon BJT Parameters

BJT Parameters

#	Name	Parameter	Unit	Default	Example	Area
1	IS	transport saturation current	A	1.0E-16	1.0E-15	*
2	BF	ideal maximum forward beta	-	100	100	
3	NF	forward current emission coefficient	-	1.0	1	
4	VAF	forward Early voltage	V	infinite	200	
5	IKF	corner for forward beta high current roll-off	A	infinite	0.01	*
6	ISE	B-E leakage saturation current	A	01.0E-13		*
7	NE	B-E leakage emission coefficient	-	1.5	2	
8	BR	ideal maximum reverse beta	-	1	0.1	
9	NR	reverse current emission coefficient	-	1	1	
10	VAR	reverse Early voltage	V	infinite	200	
11	IKR	corner for reverse beta high current roll-off	A	infinite	0.01	*
12	ISC	B-C leakage saturation current	A	0	1.0E-13	*
13	NC	B-C leakage emission coefficient	-	21.5		
14	RB	zero bias base resistance	Ohms	0	100	*
15	IRB	current where base resistance falls halfway to its min value	A	infinite	0.1	*
16	RBM	minimum base resistance at high currents	Ohms	RB	10	*
17	RE	emitter resistance	Ohms	0	1	*
18	RC	collector resistance	Ohms	0	10	*
19	CJE	B-E zero-bias depletion capacitance	F	0	2PF	*
20	VJE	B-E built-in potential	V	0.75	0.6	
21	MJE	B-E junction exponential factor	-	0.33	0.33	
22	TF	ideal forward transit time	sec	0	0.1Ns	
23	XTF	coefficient for bias dependence of TF	-	0		
24	VTF	voltage describing VBC dependence of TF	V	infinite		
25	ITF	high-current parameter for effect on TF	A	0		*
26	PTF	excess phase at freq=1.0/(TF*2PI) Hz	deg	0		
27	CJC	B-C zero-bias depletion capacitance	F	0	2PF	*
28	VJC	B-C built-in potential	V	0.75	0.5	
29	MJC	B-C junction exponential factor	-	0.33	0.5	
30	XCJC	fraction of B-C depletion capacitance connected to internal base node	-	1		
31	TR	ideal reverse transit time	sec	0	10Ns	
32	CJS	zero-bias collector-substrate capacitance	F	0	2PF	*
33	VJS	substrate junction built-in potential	V	0.75		
34	MJS	substrate junction exponential factor	-	0	0.5	
35	XTB	forward and reverse beta temperature exponent	-	0		
36	EG	energy gap for temperature effect on IS	eV	1.11		
37	XTI	temperature exponent for effect on IS	-	3		
38	KF	flicker-noise coefficient	-	0		
39	AF	flicker-noise exponent	-	1		
40	FC	coefficient for forward-bias depletion capacitance formula	-	0.5		

The dc model is defined by the parameters IS, BF, NF, ISE, IKF, and NE which determine the forward current gain characteristics, IS, BR, NR, ISC, IKR and NC which determine the reverse current gain characteristics, and VAF and VAR which determine the output conductance for forward and reverse regions. Three ohmic resistances RB, RC and RE are included, where RB can be high current dependent. Base charge storage is modeled by forward and reverse transit times TF and TR, the forward transit time TF being bias dependent if desired, and nonlinear depletion layer capacitances which are determined by CJE, VJE and MJE for the B-E junction, CJC, VJC and MJC for the B-C junction and CJS, VJS and MJS for the C-S (Collector-Substrate) junction. The temperature dependence of the saturation current, IS, is determined by the energy gap, EG, and the saturation current temperature exponent, XTI. Additionally base current temperature dependence is modeled by the beta temperature exponent XTB in the model.

8.4   DESCRIBING JUNCTION FIELD-EFFECT TRANSISTORS TO SPICE

8.4.1   JFET Element Line

The general form for an element line for a junction field-effect transistor is

    JXXXXXX ND NG NS MODNAME <AREA> <OFF> <IC=VDS,VGS>

where JXXXXXX is the name of the transistor. ND, NG and NS are the nodes to which the drain, gate and source are connected, respectively. MODNAME is the model name which is used in an associated .MODEL control line. All of these are required in the JFET element line. The optional parameters are

AREA - the area factor which determines how many of the JFET model MODNAME are put in parallel to make one JXXXXXX.

OFF - an initial condition of JXXXXXX for the DC analysis.

IC=VDS,VGS - a specification of initial conditions, for use with the UIC option of the .TRAN control line. This causes SPICE to use VDS and VGS as the initial conditions for drain-source and gate-source voltages instead of the quiescent operating point junction voltages when a transient analysis is done.

8.4.2   JFET Model Line

The general forms for model lines for junction field-effect transistors are

> .MODEL MODNAME NJF<(PAR1=PVAL1 PAR2=PVAL2 ...)>
> .MODEL MODNAME PJF<(PAR1=PVAL1 PAR2=PVAL2 ...)>

where MODNAME is the model name given to a JFET in an element line, and NJF or PJF tells SPICE that the semiconductor being modeled is an N-channel or P-channel JFET and not a diode, BJT or MOSFET. PAR is the parameter name, which can be any of those in the parameter list in Section 8.4.3, and PVAL is the value of that parameter. Note that the PAR and PVAL parts of the .MODEL line are optional; the minimum specification of a JFET is simply NJF or PJF.

There are 12 parameters that can be specified in the .MODEL control line for a JFET. The model is derived from the FET model of Shichman and Hodges.

EXAMPLE:

> JBUFFER 8 5 7 EASY
> .MODEL EASY PJF

JBUFFER has its drain, gate and source connected to nodes 8, 5 and 7, respectively. It is a P-channel JFET, which has the 12 default parameters listed in Section 8.4.3.

EXAMPLE:

> JTOP   3 8 9 JBIG 5
> JMID   1 2 4 JBIG
> JBOTTOM 6 7 9 JBIG 5
> .MODEL JBIG NJF(VTO = -3.5)

Junction field effect transistors JTOP and JBOTTOM have area factors of 5, which means their physical areas will be 5 times the area of JMID. All parameters shown in Section 8.4.3 with an asterisk (*) in the area column will be affected by the area factor. By default, JMID has an area factor of 1.

All three N-channel JFETs have a threshold voltage of -3.5 V.   Note that only one .MODEL line is needed for these three transistors.   Any number of transistors that have the same MODNAME can use the same .MODEL line.

8.4.3   JFET Model Parameters

JFET Parameters

#	Name	Parameter	Units	Default	Example	Area
1	VTO	threshold voltage	V	-2.0	-2.0	
2	BETA	transconductance parameter	A/V**2	1.0E-4	1.0E-3	*
3	LAMBDA	channel length modulation parameter	1/V	0	1.0E-4	
4	RD	drain ohmic resistance	Ohm	0	100	*
5	RS	source ohmic resistance	Ohm	0	100	*
6	CGS	zero-bias G-S junction capacitance	F	0	5PF	*
7	CGD	zero-bias G-D junction capacitance	F	0	1PF	*
8	PB	gate junction potential	V	1	0.6	
9	IS	gate junction saturation current	A	1.0E-14	1.0E-14	*
10	KF	flicker noise coefficient	-	0		
11	AF	flicker noise exponent	-	1		
12	FC	coefficient for forward-bias depletion capacitance formula	-	0.5		

The JFET model is derived from the FET model of Shichman and Hodges.   The dc characteristics are defined by the parameters VTO and BETA, which determine the variation of drain current with gate voltage, LAMBDA, which determines the output conductance, and IS, the saturation current of the two gate junctions.   Two ohmic resistances, RD and RS, are included. Charge storage is modeled by nonlinear depletion layer capacitances for both gate junctions which vary as the -1/2 power of junction voltage and are defined by the parameters CGS, CGD and PB.

8.5   DESCRIBING MOS FIELD–EFFECT TRANSISTORS TO SPICE

8.5.1   MOSFET Element Line

The general form for an element line for a metal-oxide semiconductor field-effect transistor is

```
MXXXXXX ND NG NS NB MODNAME <L=VAL> <W=VAL>
+ <AD=VAL> <AS=VAL> <PD=VAL> <PS=VAL> <NRD=VAL>
+ <NRS=VAL>  <OFF>  <IC=VDS,VGS,VBS>
```

where MXXXXXX is the name of the MOSFET transistor. ND, NG, NS and NB are the nodes to which the drain, gate, source and bulk (or substrate) are connected, respectively. MODNAME is the model name which is used in an associated .MODEL control line. All of these are required in the MOSFET element line. The optional parameters and their meanings are

L	length of the channel, in meters
W	width of the channel, in meters
AD	area of drain diffusion, in square meters
AS	area of source diffusion, in square meters

If any of the above parameters are not specified, default values are used. The user may specify the values to be used for these default parameters on the .OPTIONS control line (not to be confused with the .MODEL control line). The parameters on the .OPTIONS control line are DEFL for L, DEFW for W, DEFAD for AD, and DEFAS for AS. Use of defaults done this way simplifies the writing of the input file initially and subsequent file editing that would occur if device geometries are changed. See Appendix A.3 for the default values.

The rest of the optional parameters and their meanings are:

PD	perimeter of drain junction, in meters
PS	perimeter of source junction, in meters
NRD	equivalent number of squares of drain diffusion
NRS	equivalent number of squares of source diffusion; NRD and NRS values multiply the sheet resistance RSH specified on the .MODEL line for an accurate representation of the parasitic series drain and source resistance of each transistor. PD and PS default to 0.0 while NRD and NRS default to 1.0
OFF	indicates an initial condition on the device for DC analysis
IC=	sets the initial conditions for drain-source voltage, gate-source voltage and substrate-source voltage. This is intended to be used with the UIC option of the .TRAN control line. This causes SPICE to use the VDS, VGS and VBS values instead of the quiescent operating point voltages when a transient analysis is done

8.5.2   MOSFET Model Line

The general forms for model lines for metal-oxide semiconductor field-effect transistors are

> .MODEL MODNAME NMOS<(PAR1=PVAL1 PAR2=PVAL2 ...)>
> .MODEL MODNAME PMOS<(PAR1=PVAL1 PAR2=PVAL2 ...)>

where MODNAME is the model name given to a MOSFET in an element line, and NMOS or PMOS tells SPICE that the semiconductor being modeled is an N-channel or P-channel MOSFET and not a diode, BJT or JFET.   PAR is the parameter name, which can be any of those in the parameter list in Section 8.5.3, and PVAL is the value of that parameter.   Note that the PAR and PVAL parts of the .MODEL line are optional; the minimum specification of a MOSFET is simply NMOS or PMOS.

There are 42 parameters that can be specified in the .MODEL control line for a MOSFET.

EXAMPLE:

> MBUFFER 8 5 7 0 SIMPLE
> .MODEL SIMPLE NMOS

Transistor MBUFFER has its drain, gate, source and substrate connected to nodes 8, 5, 7 and 0 respectively.   It is an N-channel MOSFET, which has the 42 default parameters listed in Section 8.4.3.   It also has geometry information which is either specified in the .OPTIONS control line or will default to those values listed in the .OPTIONS section.   Those default values are: DEFL = 100 microns, DEFW = 100 microns, DEFAD = 0 square meters and DEFAS = 0 square meters.

EXAMPLE:

> MTOP   3 8 9 5 TYPEA 80U
> MMID   1 2 4 5 TYPEA 60U
> MBOTTOM 6 7 9 5 TYPEA
> .MODEL TYPEA PMOS(KP = 2.5E-5)

Metal-oxide semiconductor field-effect transistors MTOP and MMID have channel lengths of 80 and 60 microns, respectively.   MBOTTOM uses the default value in the .OPTIONS control line.

All three P-Channel MOSFETs have a transconductance parameter of 25 microamp/(volt squared).   Note that only one .MODEL line is needed for these three transistors.   Any number of transistors that have the same MODNAME can use the same .MODEL line.

8.5.3 MOSFET Model Parameters

MOSFET Parameters

#	Name	Parameter	Units	Default	Example
1	LEVEL	model index	-	1	
2	VTO	zero-bias threshold voltage	V	0.0	1.0
3	KP	transconductance parameter	$A/V^{**}2$	2.0E-5	3.1E-5
4	GAMMA	bulk threshold parameter	$V^{**}0.5$	0.0	0.37
5	PHI	surface potential	V	0.6	0.65
6	LAMBDA	channel-length modulation (MOS1 and MOS2 only)	1/V	0.0	0.02
7	RD	drain ohmic resistance	Ohm	0.0	1.0
8	RS	source ohmic resistance	Ohm	0.0	1.0
9	CBD	zero-bias B-D junction capacitance	F	0.0	20FF
10	CBS	zero-bias B-S junction capacitance	F	0.0	20FF
11	IS	bulk junction saturation current	A	1.0E-14	1.0E-15
12	PB	bulk junction potential	V	0.8	0.87
13	CGSO	gate-source overlap capacitance per meter channel width	F/m	0.0	4.0E-11
14	CGDO	gate-drain overlap capacitance per meter channel width	F/m	0.0	4.0E-11
15	CGBO	gate-bulk overlap capacitance per meter channel length	F/m	0.0	2.0E-10
16	RSH	drain and source diffusion sheet resisitance	Ohm/sq.	0.0	10.0
17	CJ	zero-bias bulk junction bottom cap. per sq-meter of junction area	$F/m^{**}2$	0.0	2.0E-4
18	MJ	bulk junction bottom grading coef.	-	0.5	0.5
19	CJSW	zero-bias bulk junction sidewall cap. per meter of junction perimeter	F/m	0.0	1.0E-9
20	MJSW	bulk junction sidewall grading coef.	-		0.3 3
21	JS	bulk junction saturation current per sq-meter of junction area	$A/m^{**}2$	1.0E-8	
22	TOX	oxide thickness	meter	1.0E-7	1.0E-7
23	NSUB	substrate doping	$1/cm^{**}3$	0.0	4.0E15
24	NSS	surface state density	$1/cm^{**}2$	0.0	1.0E10
25	NFS	fast surface state density	$1/cm^{**}2$	0.0	1.0E10
26	TPG	type of gate material: +1 opp. to substrate -1 same as substrate 0 Al gate	-	1.0	
27	XJ	metallurgical junction depth	meter	0.0	1U
28	LD	lateral diffusion	meter	0.0	0.8U
29	UO	surface mobility	$cm^{**}2/V\text{-}s$	600	700
30	UCRIT	critical field for mobility degradation (MOS2 only)	V/cm	1.0E4	1.0E4
31	UEXP	critical field exponent in mobility degradation (MOS2 only)	-	0.0	0.1
32	UTRA	transverse field coef (mobility) (deleted for MOS2)	-	0.0	0.3
33	VMAX	maximum drift velocity of carriers	m/s	0.0	5.0E4
34	NEFF	total channel charge (fixed and mobile) coefficient (MOS2 only)	-	1.0	5.0
35	XQC	thin-oxide capacitance model flag and coefficient of channel charge share attributed to drain (0-0.5)	-	1.0	0.4
36	KF	flicker noise coefficient	-	0.0	1.0E-26
37	AF	flicker noise exponent	-	1.0	1.2
38	FC	coefficient for forward-bias depletion capacitance formula	-	0.5	
39	DELTA	width effect on threshold voltage (MOS2 and MOS3)	-	0.0	1.0
40	THETA	mobility modulation (MOS3 only)	1/V	0.0	0.1
41	ETA	static feedback (MOS3 only)	-	0.0	1.0
42	KAPPA	saturation field factor (MOS3 only)	-	0.2	0.5

SPICE provides three MOSFET device models which differ in the formulation of the I-V characteristic. The variable LEVEL specifies the model to be used

LEVEL=1 ->   Shichman-Hodges
LEVEL=2 ->   MOS2 (as described in [1])
LEVEL=3 ->   MOS3, a semi-empirical model(see [1])

The dc characteristics of the MOSFET are defined by the device parameters VTO, KP, LAMBDA, PHI and GAMMA. These parameters are computed by SPICE if process parameters (NSUB, TOX, . . .) are given, but user-specified values always override. VTO is positive (negative) for enhancement mode and negative (positive) for depletion mode N-channel (P-channel) devices. Charge storage is modeled by three constant capacitors, CGSO, CGDO, and CGBO which represent overlap capacitances, by the nonlinear thin-oxide capacitance which is distributed among the gate, source, drain, and bulk regions, and by the nonlinear depletion-layer capacitances for both substrate junctions divided into bottom and periphery, which vary as the MJ and MJSW power of junction voltage respectively, and are determined by the parameters CBD, CBS, CJ, CJSW, MJ, MJSW and PB. There are two built-in models of the charge storage effects associated with the thin-oxide. The default is the piece-wise linear voltage-dependent capacitance model proposed by Meyer. The second choice is the charge-controlled capacitance model of Ward and Dutton [1]. The XQC model parameter acts as a flag and a coefficient at the same time. As the former it causes the program to use Meyer's model whenever larger than 0.5 or not specified, and the charge-controlled model when between 0 and 0.5. In the latter case its value defines the share of the channel charge associated with the drain terminal in the saturation region. The thin-oxide charge storage effects are treated slightly differently for the LEVEL=1 model. These voltage-dependent capacitances are included only if TOX is specified in the input description, and they are represented using Meyer's formulation.

There is some overlap among the parameters describing the junctions, e.g. the reverse current can be input either as IS (in A) or as JS (in A/m**2). Whereas the first is an absolute value the second is multiplied by AD and AS to give the reverse current of the drain and source junctions respectively. This methodology has been chosen since there is no sense in always relating junction characteristics with AD and AS entered on the device card; the areas can be defaulted. The same idea also applies to the zero-bias junction capacitances CBD and CBS (in F) on the one hand, and CJ (in F/m**2) on the other. The parasitic drain and source series resistance can be expressed as either RD and RS (in ohms) or RSH (in ohms per sq.), the latter being multiplied by the number of squares NRD and NRS input on the device card.

[1] A. Vladimirescu and S. Liu, "The Simulation of MOS Integrated Circuits Using SPICE2," ERL Memo No. ERL M80/7, Electronics Research Laboratory, University of California, Berkeley, Oct. 1980.

CHAPTER SUMMARY

SPICE has built-in models for four kinds of semiconductors: junction diode, bipolar junction transistor, junction field-effect transistor and MOS field-effect transistor.

In order to describe a semiconductor in an input file, both an element line and a .MODEL line are needed. However, many devices of the same type with identical parameters (e.g. NPN BJT with BF = 125) can use the same .MODEL line.

For many circuit simulations only a few key parameters need to be specified in the .MODEL line to SPICE. The default parameters give excellent results in a large proportion of analyses.

Chapter 9

SUBCIRCUITS

9.1 THE NEED FOR SUBCIRCUITS IN SPICE INPUT FILES

Many circuits are made with basic electronic building blocks such as operational amplifiers, logic gates and comparators. For example, an active filter might contain six identical op-amps. One way to write an input file for such an active filter for SPICE analysis would be to describe the op-amp six times. If it took 10 element lines to describe the op-amp, then for the active filter 60 element lines would be needed just for the op-amps. In addition to 60 element lines for op-amps, additional element lines would be needed for resistors and capacitors external to the op-amps. Using this approach is tedious work; fortunately, SPICE permits us to define an often-used circuit block only once in each input file. Then that circuit block can be used many times by writing a single element line each time the circuit block is used. SPICE calls such circuit blocks subcircuits.

9.2   HOW TO USE SUBCIRCUITS

9.2.1   Control Lines Used For Subcircuit Definition

In order to use a subcircuit in an input file, the subcircuit must be defined. The format of the first line of a subcircuit definition is

.SUBCKT SUBNAME N1 <N2 N3 ... >

and the last line of a subcircuit definition is

.ENDS <SUBNAME>

The .ENDS control line should not be confused with the last control line of the entire input file, .END. In a SPICE input file with three subcircuits, there would be three .ENDS SUBNAME control lines and only one .END control line.
SUBNAME is the name you will call the subcircuit. N1, N2 . . . are those nodes of the subcircuit that will be connected to other nodes in the circuit; they are called the external nodes of the subcircuit. Node zero may not be an external node of the subcircuit. In addition to the external nodes, a subcircuit can have any number of internal nodes. Internal nodes are nodes within the subcircuit definition which are not external nodes; node zero may be an internal node. Such internal nodes are local to that subcircuit (except for node zero), and are not considered by SPICE to be connected to the circuit outside of the subcircuit, even if the node numbers are the same. However, in the interests of clarity, not to mention kindness to others who may look at your input file, it is best not to use the same node numbers within a subcircuit and also in the circuit.

Allowed within a subcircuit, between the .SUBCKT and .ENDS control lines, are element lines of the subcircuit, semiconductor device models, other subcircuit definitions, and even subcircuit calls. Not allowed within a subcircuit definition are control lines.
The .ENDS line must have the optional SUBNAME included only if subcircuit definitions are nested (only if a subcircuit definition contains another subcircuit definition).

9.2.2   Element Line to Use a Subcircuit in a Circuit

XYYYYYY N1 <N2 N3 ... > SUBNAME

XYYYYYY is the element name of the pseudo-element which is defined elsewhere in the input file as SUBNAME.   N1 . . . are the circuit nodes to which the subcircuit is connected.

9.2.3   Example of Subcircuit Use: An R-C Circuit

Consider the R-C circuit fed by a sinusoidal voltage source shown in Fig. 9.1.   First we'll do an AC analysis without subcircuits.   The SPICE input file shown in Fig. 9.2 con-

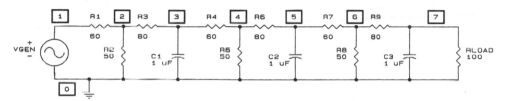

Figure 9.1   R-C Circuit with Sinusoidal
Voltage Source.

```
RC1.CIR    LONG R-C NETWORK
VGEN    1    0    AC    1
R1      1    2    60
R2      2    0    50
R3      2    3    80
C1      3    0    1U
R4      3    4    60
R5      4    0    50
R6      4    5    80
C2      5    0    1U
R7      5    6    60
R8      6    0    50
R9      6    7    80
C3      7    0    1U
RLOAD   7    0    100
.AC  DEC  5  100  1MEG
.PLOT  AC  VM(7,0)
.OPTIONS NOPAGE
.END
```

Figure 9.2   Input File RC1.CIR for R-C
Circuit.

```
********11/ 2/87******** Demo PSpice (May 1986) ******* 9:19:51********

RC1.CIR    LONG R-C NETWORK

****     CIRCUIT DESCRIPTION

***********************************************************************

VGEN    1   0   AC   1
R1      1   2   60
R2      2   0   50
R3      2   3   80
C1      3   0   1U
R4      3   4   60
R5      4   0   50
R6      4   5   80
C2      5   0   1U
R7      5   6   60
R8      6   0   50
R9      6   7   80
C3      7   0   1U
RLOAD   7   0   100
.AC  DEC  5  100  1MEG
.PLOT  AC  VM(7,0)
.OPTIONS NOPAGE
.END

****     SMALL SIGNAL BIAS SOLUTION      TEMPERATURE =    27.000 DEG C

  NODE   VOLTAGE       NODE   VOLTAGE     NODE   VOLTAGE      NODE   VOLTAGE


 (  1)     .0000      (  2)     .0000    (  3)     .0000    (  4)     .0000

 (  5)     .0000      (  6)     .0000    (  7)     .0000
```

**Figure 9.3 Output File RC1.OUT for R-C
Circuit.**

tains 13 element lines.   Figure 9.3 is the SPICE output file,
containing a graph of load voltage versus frequency.   As we
might predict, the network is a low-pass filter, where the
load voltage rolls off as frequency increases.

   By examining the circuit, we can see that the circuit combination of three resistors (60, 50 and 80 ohms) and one capacitor is repeated three times.   These four components can be
defined as a subcircuit, shown with the subcircuit definition
in Fig. 9.4.   Figure 9.5 is the same schematic diagram as

```
****    AC ANALYSIS                        TEMPERATURE =   27.000 DEG C


        FREQ     VM(7)


                     1.000D-15    1.000D-10    1.000D-05    1.000D+00    1.000D+05
                     - - - - - - - - - - - - - - - - - - - - - - - - - - - - - - -
    1.000D+02  1.043D-02 .          .            .            *       .            .
    1.585D+02  1.040D-02 .          .            .            *       .            .
    2.512D+02  1.033D-02 .          .            .            *       .            .
    3.981D+02  1.016D-02 .          .            .            *       .            .
    6.310D+02  9.750D-03 .          .            .            *       .            .
    1.000D+03  8.842D-03 .          .            .            *       .            .
    1.585D+03  7.120D-03 .          .            .           *        .            .
    2.512D+03  4.655D-03 .          .            .          *         .            .
    3.981D+03  2.297D-03 .          .            .        *           .            .
    6.310D+03  8.596D-04 .          .            .      *             .            .
    1.000D+04  2.636D-04 .          .            .     *              .            .
    1.585D+04  7.237D-05 .          .            .   *                .            .
    2.512D+04  1.887D-05 .          .            . *                  .            .
    3.981D+04  4.812D-06 .          .          *.                     .            .
    6.310D+04  1.216D-06 .          .         * .                     .            .
    1.000D+05  3.062D-07 .          .      *    .                     .            .
    1.585D+05  7.699D-08 .          .     *     .                     .            .
    2.512D+05  1.935D-08 .          .   *       .                     .            .
    3.981D+05  4.040D-09 .          *          .                      .            .
    6.310D+05  1.221D-09 . .      *            .                      .            .
    1.000D+06  3.067D-10 .    *                .                      .            .
                     - - - - - - - - - - - - - - - - - - - - - - - - - - - - - - -


        JOB CONCLUDED

        TOTAL JOB TIME            6.10
```

Figure 9.3 (Continued).

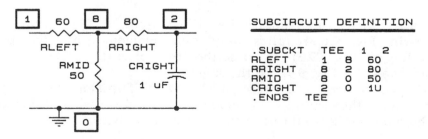

```
SUBCIRCUIT DEFINITION

.SUBCKT   TEE   1   2
RLEFT     1   8   60
RRIGHT    8   2   80
RMID      8   0   50
CRIGHT    2   0   1U
.ENDS   TEE
```

**Figure 9.4 Schematic and Definition of
Subcircuit.**

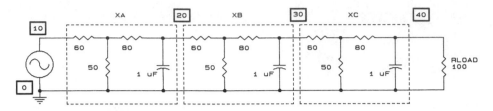

Figure 9.5 R-C Circuit with Subcircuits Shown.

```
RC2.CIR   LONG R-C DONE WITH SUBCIRCUITS
VGEN    10    0    AC    1
.SUBCKT    TEE    1    2
RLEFT    1    8    60
RMID     8    0    50
RRIGHT   8    2    80
CRIGHT   2    0    1U
.ENDS    TEE
XA       10    20    TEE
XB       20    30    TEE
XC       30    40    TEE
RLOAD    40    0    100
.AC  DEC  5  100  1MEG
.PLOT  AC  VM(40,0)
.OPTIONS  NOPAGE
.END
```

Figure 9.6 Input File RC2.CIR for R-C Circuit with Subcircuits.

Fig. 9.1, with the subcircuits illustrated.   The input file shown in Fig. 9.6 requires only nine element lines (compared with 13 in Fig. 9.2) to describe the circuit.   The analysis results, shown in Fig. 9.7, are the same.

Of course, if you were modeling a flip-flop using many nand gates, the savings in element lines by using subcircuits for the nand gates would be even more noticeable.

```
********11/ 2/87******** Demo PSpice (May 1986) ******* 9:17:48********

RC2.CIR   LONG R-C DONE WITH SUBCIRCUITS

****    CIRCUIT DESCRIPTION

**********************************************************************


VGEN    10    0    AC    1
.SUBCKT    TEE  1  2
RLEFT  1   8    60
RMID   8   0    50
RRIGHT 8   2    80
CRIGHT 2   0    1U
.ENDS  TEE
XA     10  20    TEE
XB     20  30    TEE
XC     30  40    TEE
RLOAD  40  0    100
.AC  DEC  5  100  1MEG
.PLOT  AC  VM(40,0)
.OPTIONS  NOPAGE
.END

****    SMALL SIGNAL BIAS SOLUTION     TEMPERATURE =   27.000 DEG C

  NODE   VOLTAGE    NODE   VOLTAGE    NODE   VOLTAGE    NODE   VOLTAGE


( 10)    .0000    ( 20)    .0000    ( 30)    .0000    ( 40)    .0000

( 41)    .0000    ( 42)    .0000    ( 43)    .0000
```

**Figure 9.7 Output File RC2.OUT for R-C
Circuit with Subcircuits.**

```
****    AC ANALYSIS                      TEMPERATURE =   27.000 DEG C

      FREQ     VM(40)

                   1.000D-15   1.000D-10   1.000D-05   1.000D+00   1.000D+05
                 - - - - - - - - - - - - - - - - - - - - - - - - - - - - - -
  1.000D+02   1.043D-02 .           .           .           *     .          .
  1.585D+02   1.040D-02 .           .           .           *     .          .
  2.512D+02   1.033D-02 .           .           .           *     .          .
  3.981D+02   1.016D-02 .           .           .           *     .          .
  6.310D+02   9.750D-03 .           .           .           *     .          .
  1.000D+03   8.842D-03 .           .           .           *     .          .
  1.585D+03   7.120D-03 .           .           .         *       .          .
  2.512D+03   4.655D-03 .           .           .         *       .          .
  3.981D+03   2.297D-03 .           .           .        *        .          .
  6.310D+03   8.596D-04 .           .           .       *         .          .
  1.000D+04   2.636D-04 .           .           .     *           .          .
  1.585D+04   7.237D-05 .           .           .   *             .          .
  2.512D+04   1.887D-05 .           .           .  *              .          .
  3.981D+04   4.812D-06 .           .           *.                .          .
  6.310D+04   1.216D-06 .           .         * .                 .          .
  1.000D+05   3.062D-07 .           .       *   .                 .          .
  1.585D+05   7.699D-08 .           .     *     .                 .          .
  2.512D+05   1.935D-08 .           .   *       .                 .          .
  3.981D+05   4.860D-09 .           . *         .                 .          .
  6.310D+05   1.221D-09 .          *  .         .                 .          .
  1.000D+06   3.067D-10 .        *.    .         .                .          .
                 - - - - - - - - - - - - - - - - - - - - - - - - - - - - - -

      JOB CONCLUDED

      TOTAL JOB TIME              6.40
```

Figure 9.7 (Continued).

CHAPTER SUMMARY

Subcircuits can simplify and dramatically shorten input files which use the same circuitry many times.

Subcircuits may be nested within other subcircuits. For example, a subcircuit for a shift register could have a flip-flop subcircuit within its definition, and the flip-flop subcircuit could have a nand gate subcircuit among its subcircuit element lines.

Try to use node numbers within subcircuit definitions which are different from those used outside the subcircuit. Though this is not required, doing so makes it easier for someone looking at the input file to understand.

Chapter 10

TRANSMISSION LINES

10.1 TRANSMISSION LINES ARE LOSSLESS ?

Yes, if you are using SPICE version 2, transmission lines are considered to be lossless.   This is quite reasonable for a circuit analysis program with integrated circuit emphasis, since transmission lines used within an integrated circuit (likely to be stripline or microstrip) are going to be very short, both physically and electrically.   The loss in a short transmission line is normally quite small (a fraction of a dB) and therefore negligible for most purposes.

If you wish to add loss to a SPICE transmission line, this can be done only using lumped (as opposed to distributed) R or G elements.   The dominant loss mechanism in most practical transmission lines is the conductor loss due to skin effect, not the dielectric loss.   You can simulate a long and lossy line by breaking it up into shorter sections and adding a small resistance in series with each short section to represent the distributed conductor loss.   Although it is tempting to use extremely short sections (each with a very small series R for loss), there is a tradeoff involved with this.   In transient analysis, SPICE will make the time-step (or internal computing interval) less than or equal to

one-half the minimum transmission delay of the shortest transmission line. This can make a SPICE transient analysis take
a very long time if short transmission lines are used.

Another problem is that transmission line loss increases
with frequency. It is not possible to specify a frequency-
dependent resistor, so line loss simulated by lumped resistors will be valid only at one frequency. The SPICE limitation that transmission lines are lossless is not a serious
one. In the examples that follow, you will see that SPICE
can be used to solve an assortment of transmission line problems which would be all but impossible to solve in a reasonable time without a computer.

10.2 TRANSMISSION LINE ELEMENT LINE

The two formats for an element line describing a transmission
line are

> TXXXXXX NA+ NA- NB+ NB- Z0=ZVAL F=FREQ
> + <NL=NLENGTH> <IC = VA, IA, VB, IB>

or

> TYYYYYY NA+ NA- NB+ NB- Z0=ZVAL TD=TVALUE
> + <IC = VA, IA, VB, IB>

TXXXXXX is the element name of the transmission line. NA+
and NA- are, respectively, the positive and negative nodes of
the A side of the transmission line. NB+ and NB- are, respectively, the positive and negative nodes of the B side of the
transmission line. Z0 is the characteristic impedance of the
transmission line, in ohms. (The 0 in Z0 is the number 0, not
the capital letter O.)

SPICE must be told the length of a transmission line.
There are two ways to do this; either one is acceptable, and
you can easily convert from one to the other. The two ways
are

> 1. Specify frequency and electrical length: FREQ is the
> frequency at which the transmission line is NLENGTH wave-
> lengths long. The NLENGTH parameter is dimensionless,
> since it is the normalized electrical length of the
> line. NLENGTH is the physical length of the line (in m)
> divided by the wavelength in the line (in m). The wave-
> length in the line is the free-space wavelength multi-

plied by the velocity factor of the line. If the optional NL = NLENGTH parameter is omitted, SPICE assumes a value of 0.25. That is, if the length of a transmission line is not specified, SPICE will consider it to be a one quarter wavelength section.

2. Specify the transmission delay: TD is the transmission delay of the transmission line, measured in seconds.

With each of the methods for specifying the electrical length of a transmission line you can optionally specify initial conditions at each end of the line. VA and IA are the voltage across and current into the A side of the line, while VB and IB are the voltage across and current into the B side of the line. A positive voltage means that the V+ node is more positive than the V- node on that side, while a positive current means that current is flowing into the V+ node and out of the V- node. It is important to note that these initial conditions will have an effect only if the .TRAN control line contains the option UIC (use initial conditions).

A single transmission line element line models only one propagating mode. In those circuits where the four nodes are distinct, two modes may occur: center conductor to shield and shield to ground. In order to model such a situation, two transmission line element lines should be used.

10.3 EQUIVALENCY BETWEEN TRANSMISSION DELAY AND ELECTRICAL LENGTH

Since there are two methods for telling SPICE the length of a transmission line, it may be of use to look at how each method works and how to convert from one method to the other.

In general, we can say that the transmission delay, TD, of a transmission line can be found from

TD = physical length/velocity

where velocity is the phase velocity in the line, given by

velocity = free-space velocity(velocity factor)

The velocity factor is less than or equal to 1.00. Also, the normalized electrical length, NL, is

NL = physical length/wavelength

where wavelength is the wavelength in the line, determined by

wavelength = velocity/frequency

Normalized electrical length can then be expressed as

NL = physical length (frequency)/velocity

Since physical length = TD(velocity) = NL(velocity)/frequency then

TD = NL/frequency, and NL = frequency(TD)

EXAMPLES:

1. A transmission line with a characteristic impedance of 50 ohms is 12 m long, with a velocity factor of 0.66.   a. What is the transmission delay?   b. What is the normalized electrical length?

  a. velocity = velocity factor (c), and c = 3E8 m/s, so
     velocity = 0.66 (3E8 m/s) = 2E8 m/s.
     TD = length/velocity = 12 m/(2E8 m/s) = 60 ns.

  b. Before the normalized electrical length can be determined, a frequency must be specified. Let's use 30 MHz.
     NL = frequency(length)/velocity = 30E6(12)/2E8, so NL = 1.8 wavelengths.

2. Write the element line for the above transmission line two different ways. Call it TSTUB.

  TSTUB 1 2 3 4 Z0 = 50 TD = 60N

  TSTUB 1 2 3 4 Z0 = 50 F = 30MEG NL = 1.8

  The element lines above assume TSTUB is connected to nodes 1 and 2 at the A side, and to nodes 3 and 4 at the B side.

10.4 SAMPLE TRANSMISSION LINE PROBLEMS

10.4.1 Single Pulse into Transmission Line

This problem will show how SPICE can be used in the time do-
main to look at a single short pulse at the sending end and
the receiving end of a physically short transmission line.

Figure 10.1 shows a pulse voltage source, with 50 ohm
output impedance, connected to a 50 ohm transmission line
which is terminated in a 50 ohm matched load. The SPICE
input file TLINE1.CIR, Fig. 10.2, specifies that the pulse
goes from 0 V to 10 V after a 1 ns delay from time-zero, and
has a pulse width of 3 ns.

The transmission line, with element name TSHORT, has a
transmission delay of 2 ns. We should therefore expect that
whatever voltage changes appear at the input to the transmission line (node 2) should occur 2 ns later at the output
(node 3). The input file .PLOT control line tells SPICE to
make a graph of the voltages at nodes 2 and 3 versus time,
from t = 0 to 7 ns.

The SPICE output file, Fig. 10.3, contains an element
node table showing the nodes and all elements connected to

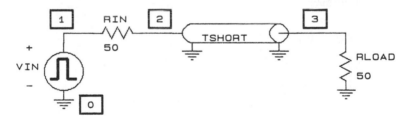

Figure 10.1 Pulse into Transmission Line.

```
TLINE1.CIR  PULSE INTO TRANSMISSION LINE
*        TO ILLUSTRATE THE TIME DELAY OF A LINE
VIN  1  0  PULSE(0 10 1N 0 0 3N)
RIN  1  2  50
TSHORT 2  0  3  0  Z0=50  TD = 2N
RLOAD 3  0  50
.TRAN .2N  7N
.PLOT TRAN V(2) (0,5) V(3)
.OPTIONS NOPAGE NODE
.END
```

Figure 10.2 Input File TLINE1.CIR.

```
********11/22/87******** Demo PSpice (May 1986) ********19:54:40********


TLINE1.CIR  PULSE INTO TRANSMISSION LINE


****    CIRCUIT DESCRIPTION


***********************************************************************


*       TO ILLUSTRATE THE TIME DELAY OF A LINE
VIN  1   0   PULSE(0 10 1N 0 0 3N)
RIN  1  2  50
TSHORT 2  0  3  0  Z0=50  TD = 2N
RLOAD 3  0  50
.TRAN .2N  7N
.PLOT TRAN V(2)  (0,5)  V(3)
.OPTIONS NOPAGE NODE
.END

****    ELEMENT NODE TABLE


      0    RLOAD    VIN      TSHORT    TSHORT

      1    RIN      VIN

      2    RIN      TSHORT

      3    RLOAD    TSHORT

****    INITIAL TRANSIENT SOLUTION      TEMPERATURE =   27.000 DEG C

  NODE   VOLTAGE     NODE   VOLTAGE    NODE   VOLTAGE


( 1)    .0000    ( 2)    .0000   ( 3)    .0000
```

Figure 10.3 Output File TLINE1.OUT.

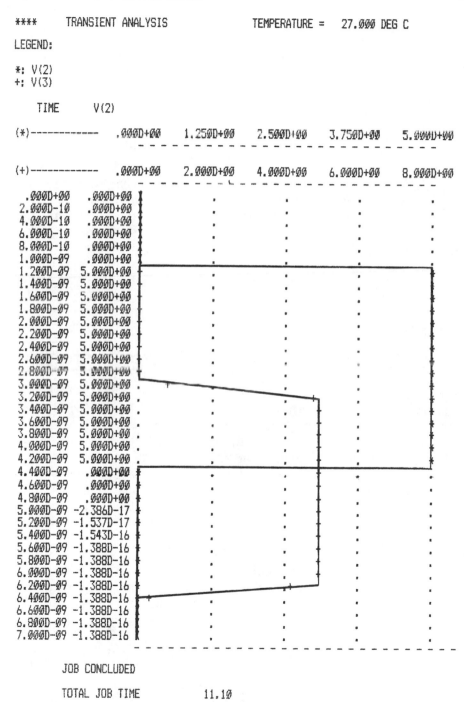

```
****      TRANSIENT ANALYSIS              TEMPERATURE =   27.000 DEG C

LEGEND:

*: V(2)
+: V(3)

     TIME      V(2)

(*)------------    .000D+00    1.250D+00    2.500D+00    3.750D+00    5.000D+00

                                - - - - - - - - - - - - - - - - - - - - - - - - -

(+)------------    .000D+00    2.000D+00    4.000D+00    6.000D+00    8.000D+00
                                - - - - - - - - - - - - - - - - - - - - - - - - -
  .000D+00    .000D+00    *
 2.000D-10    .000D+00    *
 4.000D-10    .000D+00    *
 6.000D-10    .000D+00    *
 8.000D-10    .000D+00    *
 1.000D-09    .000D+00    *
 1.200D-09   5.000D+00
 1.400D-09   5.000D+00
 1.600D-09   5.000D+00
 1.800D-09   5.000D+00
 2.000D-09   5.000D+00
 2.200D-09   5.000D+00
 2.400D-09   5.000D+00
 2.600D-09   5.000D+00
 2.800D-09   5.000D+00
 3.000D-09   5.000D+00
 3.200D-09   5.000D+00
 3.400D-09   5.000D+00
 3.600D-09   5.000D+00
 3.800D-09   5.000D+00
 4.000D-09   5.000D+00
 4.200D-09   5.000D+00
 4.400D-09    .000D+00
 4.600D-09    .000D+00
 4.800D-09    .000D+00
 5.000D-09  -2.386D-17
 5.200D-09  -1.537D-17
 5.400D-09  -1.543D-16
 5.600D-09  -1.388D-16
 5.800D-09  -1.388D-16
 6.000D-09  -1.388D-16
 6.200D-09  -1.388D-16
 6.400D-09  -1.388D-16
 6.600D-09  -1.388D-16
 6.800D-09  -1.388D-16
 7.000D-09  -1.388D-16

      JOB CONCLUDED

      TOTAL JOB TIME          11.10
```

Figure 10.3 (Continued).

those nodes, and a graph of the transient analysis results. Voltage V(2) is plotted with asterisks (*), and V(3) is plotted with pluses (+). Notice that the scales for the two plots are different, as specified in the .PLOT control line. V(2) plot limits are 0 and 5 V, while SPICE automatically scaled V(3) from 0 to 8V.

The input voltage, V(2), rises from 0 V to 5 V at 1 ns and remains 5 V for 3 ns. The output voltage, V(3), does the same as V(2) except that it is delayed by 2 ns. Where the two graphs overlap, a letter X is used for the plot symbol. Since the transmission line is terminated in a matched load, no reflections occur. If the load is not matched, a reflection will occur at the mismatch, and the input voltage will change at a time equal to twice the transmission delay of the line. This is the basis of time-domain reflectometry (TDR) which is used to examine transmission lines for defects from the input side. SPICE can be used to predict in advance what a TDR display should be for a certain transmission line circuit.

10.4.2 Transmission Line Balun Power Bandwidth

SPICE will be used to determine the power bandwidth of a transmission line balanced-unbalanced (balun) transformer. Baluns are used to connect coaxial transmission lines to balanced transmission lines or antennas, and utilize a one-half wavelength section of transmission line.

It is a very simple task to analyze how baluns work at the design frequency, where the line is indeed one-half wavelength long. However, it is not at all simple to analyze how the balun will perform at frequencies above or below the design frequency. SPICE can be used to great advantage to perform a frequency sweep of the input voltage, and to plot the output voltage versus frequency. We can tell the power bandwidth of the balun circuit by observing when the output voltage magnitude drops to 0.707 of its value at the design frequency.

The circuit is shown in Fig. 10.4. The 75 ohm coaxial transmission line impedance is converted by the balun to the 300 ohm impedance of the load. The design frequency of the balun is 100 MHz, so the transmission line T2 is 0.5 wavelength long at 100 MHz. Figure 10.5 shows the input file, BALUN.CIR, in which the AC analysis control line sweeps the frequency linearly from 20 MHz to 180 MHz.

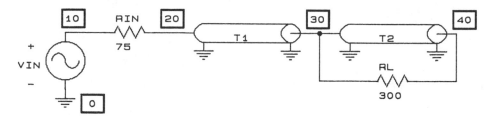

Figure 10.4 Transmission Line Balun Circuit.

The output file (Fig. 10.6) shows the graph of the volt-age magnitude across the load resistor, VM(30,40), to be a maximum of 20 V at 100 MHz. Using the 0.707 criterion, we can see that the voltage falls to 0.707(20 V), or 14.14 V at about 48 MHz and 152 MHz. Thus, SPICE has directly provided the information to determine the power bandwidth to be about 152 - 48, or 104 MHz. The power bandwidth is not necessarily the same as the usable bandwidth, since the voltage standing-wave ratio, or VSWR, may be acceptable only over a much nar-rower bandwidth. The higher the VSWR, the more power is re-flected by the load back to the source. The next problem will show how to make SPICE provide information which can be used to calculate VSWR at each frequency.

10.4.3 Transmission Line Balun Input Impedance

A very simple modification can be made to a circuit to make SPICE print out the input impedance of a circuit. In this problem we will slightly change the circuit of the previous problem and cause SPICE to print out the input impedance of the balun circuit at each frequency. From this input impe-dance data the VSWR at each frequency can easily be calcula-ted, and a VSWR-based bandwidth can be determined.

```
BALUN.CIR    TRANS. LINE BALUN
VIN    10  0  AC  20
RIN  10  20   75
T1    20  0  30  0   Z0=75  F=100MEG NL=1.0
T2    30  0  40  0   Z0=75  F=100MEG NL=0.5
RL   30  40  300
.AC LIN 41  20MEG 180MEG
.PLOT AC VM(30,40) (3,20)
.OPTIONS NOPAGE NODE
.END
```

Figure 10.5 Input File BALUN.CIR.

```
********11/ 5/87******** Demo PSpice (May 1986) ******* 8:49:59********


BALUN.CIR   TRANS. LINE BALUN


****     CIRCUIT DESCRIPTION


**************************************************************************


VIN    10  0  AC  20
RIN  10  20  75
T1    20  0  30  0  Z0=75  F=100MEG NL=1.0
T2    30  0  40  0  Z0=75  F=100MEG NL=0.5
RL   30  40  300
.AC LIN 41  20MEG 180MEG
.PLOT AC VM(30,40) (3,20)
.OPTIONS NOPAGE NODE
.END

****     ELEMENT NODE TABLE


     0    VIN       T1       T1       T2       T2

    10    RIN       VIN

    20    RIN       T1

    30    RL        T1       T2

    40    RL        T2

****     SMALL SIGNAL BIAS SOLUTION      TEMPERATURE =   27.000 DEG C

 NODE   VOLTAGE     NODE   VOLTAGE     NODE   VOLTAGE     NODE   VOLTAGE

 ( 10)    .0000   ( 20)    .0000   ( 30)    .0000   ( 40)    .0000
```

Figure 10.6 Output File BALUN.OUT.

```
****     AC ANALYSIS                        TEMPERATURE =   27.000 DEG C


     FREQ      VM(30,40)

                    3.000D+00    4.821D+00    7.746D+00    1.245D+01    2.000D+01
                    - - - - - - - - - - - - - - - - - - - - - - - - - - - - - -
2.000D+07  3.729D+00 .     *       .          .          .          .
2.400D+07  5.205D+00 .             . *        .          .          .
2.800D+07  6.805D+00 .             .          *          .          .
3.200D+07  8.453D+00 .             .          . *        .          .
3.600D+07  1.008D+01 .             .          .          *          .
4.000D+07  1.162D+01 .             .          .          . *        .
4.400D+07  1.303D+01 .             .          .          .  *       .
4.800D+07  1.429D+01 .             .          .          .     *    .
5.200D+07  1.538D+01 .             .          .          .       *  .
5.600D+07  1.632D+01 .             .          .          .        * .
6.000D+07  1.711D+01 .             .          .          .         *.
6.400D+07  1.777D+01 .             .          .          .          *
6.800D+07  1.831D+01 .             .          .          .          * .
7.200D+07  1.876D+01 .             .          .          .          * .
7.600D+07  1.912D+01 .             .          .          .          *.
8.000D+07  1.941D+01 .             .          .          .          *.
8.400D+07  1.963D+01 .             .          .          .          *.
8.800D+07  1.980D+01 .             .          .          .          *
9.200D+07  1.991D+01 .             .          .          .          *
9.600D+07  1.998D+01 .             .          .          .          *
1.000D+00  2.000D+01 .             .          .          .          *
1.040D+08  1.998D+01 .             .          .          .          *
1.080D+08  1.991D+01 .             .          .          .          *
1.120D+08  1.980D+01 .             .          .          .          *
1.160D+08  1.963D+01 .             .          .          .          *.
1.200D+08  1.941D+01 .             .          .          .          *.
1.240D+08  1.912D+01 .             .          .          .          *.
1.280D+08  1.876D+01 .             .          .          .          * .
1.320D+08  1.831D+01 .             .          .          .          * .
1.360D+08  1.777D+01 .             .          .          .         *
1.400D+08  1.711D+01 .             .          .          .        *
1.440D+08  1.632D+01 .             .          .          .       *  .
1.480D+08  1.538D+01 .             .          .          .     *    .
1.520D+08  1.429D+01 .             .          .          .   *      .
1.560D+08  1.303D+01 .             .          .          . *        .
1.600D+08  1.162D+01 .             .          .          *          .
1.640D+08  1.008D+01 .             .          .        *            .
1.680D+08  8.453D+00 .             .          . *        .          .
1.720D+08  6.805D+00 .             .          *          .          .
1.760D+08  5.205D+00 .             . *        .          .          .
1.800D+08  3.729D+00 .     *       .          .          .          .
                    - - - - - - - - - - - - - - - - - - - - - - - - - - - - - -


     JOB CONCLUDED

     TOTAL JOB TIME          9.40
```

Figure 10.6 (Continued).

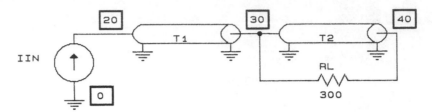

Figure 10.7 Transmission Line Balun Circuit, Current Input.

If an AC current source of 1 A magnitude and 0 degrees phase is connected to the input of a circuit, then the phasor voltage at the input is the product of current and impedance:

$$V = I\,(Z) = 1\,(Z) = Z$$

Thus, by replacing the AC voltage source and 75 ohm resistor in the previous problem with a 1 A current source, SPICE will generate input impedance data. Refer to the modified schematic diagram in Fig. 10.7 which shows current source IIN. Figure 10.8 shows the input file, BALUN2.CIR, for this circuit. Since the phase is not specified in the IIN element line, the default value of 0 degrees will apply. Notice that the + node for IIN is 0, and the - node is 20. That means that positive current flows up from ground (node 0) through IIN and into node 20.

The output file shown in Fig. 10.9 has a table of the complex input voltage (and impedance) in rectangular form,

```
BALUN2.CIR   TRANS. LINE BALUN, 1A CURRENT INPUT
IIN    0  20  AC   1
T1    20  0  30  0  Z0=75  F=100MEG NL=1.0
T2    30  0  40  0  Z0=75  F=100MEG NL=0.5
RL    30  40  300
.AC LIN 41  60MEG 140MEG
.PLOT AC VM(20) VP(20)
.PRINT AC VR(20) VI(20)
.OPTIONS NOPAGE NODE
.END
```

Figure 10.8 Input File BALUN2.CIR.

```
********11/ 5/87******** Demo PSpice (May 1986) ******* 8:15:30********

BALUN2.CIR    TRANS. LINE BALUN, 1A CURRENT INPUT

****    CIRCUIT DESCRIPTION

***********************************************************************


IIN   0  20  AC  1
T1    20  0   30  0   Z0=75  F=100MEG NL=1.0
T2    30  0   40  0   Z0=75  F=100MEG NL=0.5
RL    30  40  300
.AC LIN 41  60MEG 140MEG
.PLOT AC VM(20) VP(20)
.PRINT AC VR(20) VI(20)
.OPTIONS NOPAGE NODE
.END

****    ELEMENT NODE TABLE


      0    IIN      T1       T1       T2       T2

     20    IIN      T1

     30    RL       T1       T2

     40    RL       T2

****    SMALL SIGNAL BIAS SOLUTION       TEMPERATURE =    27.000 DEG C

 NODE   VOLTAGE      NODE   VOLTAGE      NODE   VOLTAGE


( 20)    .0000    ( 30)     .0000    ( 40)     .0000
```

Figure 10.9 Output File BALUN2.OUT.

```
****     AC ANALYSIS                    TEMPERATURE =    27.000 DEG C


     FREQ        VR(20)       VI(20)


   6.000E+07    4.049E+01    5.712E+01
   6.200E+07    5.716E+01    7.107E+01
   6.400E+07    8.338E+01    8.131E+01
   6.600E+07    1.202E+02    7.845E+01
   6.800E+07    1.557E+02    4.991E+01
   7.000E+07    1.647E+02    2.408E+00
   7.200E+07    1.438E+02   -3.412E+01
   7.400E+07    1.153E+02   -4.819E+01
   7.600E+07    9.230E+01   -4.809E+01
   7.800E+07    7.653E+01   -4.205E+01
   8.000E+07    6.640E+01   -3.413E+01
   8.200E+07    6.028E+01   -2.597E+01
   8.400E+07    5.706E+01   -1.821E+01
   8.600E+07    5.605E+01   -1.113E+01
   8.800E+07    5.682E+01   -4.924E+00
   9.000E+07    5.910E+01    1.576E-01
   9.200E+07    6.254E+01    3.799E+00
   9.400E+07    6.668E+01    5.640E+00
   9.600E+07    7.078E+01    5.435E+00
   9.800E+07    7.385E+01    3.315E+00
   1.000E+08    7.500E+01    1.818E-13
   1.020E+08    7.385E+01   -3.315E+00
   1.040E+08    7.078E+01   -5.435E+00
   1.060E+08    6.668E+01   -5.640E+00
   1.080E+08    6.254E+01   -3.799E+00
   1.100E+08    5.910E+01   -1.576E-01
   1.120E+08    5.682E+01    4.924E+00
   1.140E+08    5.605E+01    1.113E+01
   1.160E+08    5.706E+01    1.821E+01
   1.180E+08    6.028E+01    2.597E+01
   1.200E+08    6.640E+01    3.413E+01
   1.220E+08    7.653E+01    4.205E+01
   1.240E+08    9.230E+01    4.809E+01
   1.260E+08    1.153E+02    4.819E+01
   1.280E+08    1.438E+02    3.412E+01
   1.300E+08    1.647E+02   -2.408E+00
   1.320E+08    1.557E+02   -4.991E+01
   1.340E+08    1.202E+02   -7.845E+01
   1.360E+08    8.338E+01   -8.131E+01
   1.380E+08    5.716E+01   -7.107E+01
   1.400E+08    4.049E+01   -5.712E+01
```

Figure 10.9 (Continued).

```
****     AC ANALYSIS                          TEMPERATURE =   27.000 DEG C
LEGEND:
*: VM(20)
+: VP(20)
    FREQ      VM(20)
(*)------------- 3.981D+01     6.310D+01     1.000D+02     1.585D+02     2.512D+02
                           - - - - - - - - - - - - - - - - - - - - - - - - - -
(+)------------- -1.000D+02   -5.000D+01     .000D+00     5.000D+01     1.000D+02
                           - - - - - - - - - - - - - - - - - - - - - - - - - -
 6.000D+07  7.002D+01 .              .     *    .            .+         .
 6.200D+07  9.121D+01 .              .         *  .          .+         .
 6.400D+07  1.165D+02 .              .            .    *     .+,        .
 6.600D+07  1.435D+02 .              .            .      +*  .          .
 6.800D+07  1.635D+02 .              .            .    +    .*          .
 7.000D+07  1.647D+02 .              .            .  +      .*          .
 7.200D+07  1.478D+02 .              .          +      .  *. .          .
 7.400D+07  1.250D+02 .              .      +          .  *  .          .
 7.600D+07  1.041D+02 .              .       +     .*   .    .          .
 7.800D+07  8.732D+01 .              .       + *   .    .    .          .
 8.000D+07  7.466D+01 .              .   X+  .     .    .    .          .
 8.200D+07  6.564D+01 .              . .* +  .     .    .    .          .
 8.400D+07  5.989D+01 .            *.      + .     .    .    .          .
 8.600D+07  5.714D+01 .      *     .      + .      .    .    .          .
 8.800D+07  5.704D+01 .      * .          +,       .    .    .          .
 9.000D+07  5.910D+01 .      * .           +       .    .    .          .
 9.200D+07  6.266D+01 .         *          ,+      .    .    .          .
 9.400D+07  6.692D+01 .          *        ..+      .    .    .          .
 9.600D+07  7.099D+01 .      .  *         .+       .    .    .          .
 9.800D+07  7.392D+01 .      .    *       .+       .    .    .          .
 1.000D+08  7.500D+01 .      .      *     +        .    .    .          .
 1.020D+08  7.392D+01 .      .    *       +,       .    .    .          .
 1.040D+08  7.099D+01 .      .  *         +,       .    .    .          .
 1.060D+08  6.692D+01 .      . *          +,       .    .    .          .
 1.080D+08  6.266D+01 .         *         +,       .    .    .          .
 1.100D+08  5.910D+01 .      * .          +        .    .    .          .
 1.120D+08  5.704D+01 .      * .          .+       .    .    .          .
 1.140D+08  5.714D+01 .      * .            +      .    .    .          .
 1.160D+08  5.989D+01 .      *.             +      .    .    .          .
 1.180D+08  6.564D+01 .       .*            +      .    .    .          .
 1.200D+08  7.466D+01 .          *          +     .    .    .          .
 1.220D+08  8.732D+01 .              *      +     .    .    .          .
 1.240D+08  1.041D+02 .              .*     +     .    .    .          .
 1.260D+08  1.250D+02 .              .    X       .    .    .          .
 1.280D+08  1.478D+02 .              .    +  *.   .    .    .          .
 1.300D+08  1.647D+02 .              .  +      .*  .    .    .          .
 1.320D+08  1.635D+02 .          +          .*  .    .    .          .
 1.340D+08  1.435D+02 .          +          .    *  .    .          .
 1.360D+08  1.165D+02 .      .+             .    *     .    .          .
 1.380D+08  9.121D+01 .      +          *          .    .          .
 1.400D+08  7.002D+01 .      +.  *        .          .    .          .
                           - - - - - - - - - - - - - - - - - - - - - - - - - -
          JOB CONCLUDED
          TOTAL JOB TIME        10.70
```

Figure 10.9 (Continued).

VR(20) and VI(20), as a function of frequency. Also a graph of the complex input impedance in polar form, VM(20) and VP(20), versus frequency is plotted.

The input impedance data, when converted into VSWR data, indicates that the VSWR at 60 MHz and 140 MHz exceeds 3. The conversion method can be found in many texts on electronic communications with a chapter on transmission lines. While the data at 48 MHz and 152 MHz was not printed in this analysis, the VSWR at those frequencies is 5.66. A VSWR that high would be unacceptable for most applications.

10.4.4 Impedance Match with Quarter Wavelength Transmission Line Transformer

One method of matching a load to a transmitter is to use a quarter-wavelength section of transmission line to transform the impedance at its load side to the characteristic impedance of the system. Figure 10.10 shows a circuit where transmission line T2 is a quarter-wavelength long and transforms the impedance of 500 ohms at node 4 to 50 ohms at node 3. The purpose of T3 is to transform the very low resistance of the load, 5 ohms, to 500 ohms at node 4.

The input file in Fig. 10.11, MATCH.CIR, has element lines to describe the circuit, and calls for an AC analysis. The results of the AC analysis will be printed and plotted. The dead voltage source VSENSE is used to measure the current

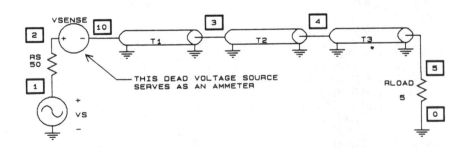

Figure 10.10 Quarterwave Transformer Transmission Line Match Circuit.

```
MATCH.CIR    0.25 WAVELENGTH TRANS. LINE MATCH
VS      1   0   AC   141.42
RS      1   2   50
VSENSE  2   10  0
T1      10  0   3   0   Z0=50     F=50MEG  NL=2
T2      3   0   4   0   Z0=158.11 F=50MEG  NL=0.25
*       SQRT(500*50) = 158.11, 0.25 WAVE MATCH
T3      4   0   5   0   Z0=50     F=50MEG  NL=0.25
*       T3 SECTION TRANSFORMS 5 OHM RLOAD TO 500 OHM
RLOAD   5   0   5
.AC LIN 41  10MEG  90MEG
.PRINT AC VM(10)  VP(10)  IM(VSENSE)  IP(VSENSE)  VM(5)
.PLOT AC VM(5) (5,25)
.PLOT AC VDB(5) (16,28)
.OPTIONS NOPAGE NODE
.END
```

Figure 10.11   Input File MATCH.CIR.

flowing into the circuit. The voltage magnitude across the load, V(5), is plotted two ways. The first .PLOT control line plots the magnitude of node 5 voltage; the second plots 20*LOG(node 5 voltage). The decibel data that results makes it easy to evaluate the half power, or 3 dB, bandwidth of this matching circuit.

The output file in Fig. 10.12 contains both a table of polar form input voltage and current, load voltage magnitude, and also two graphs of load voltage. The VDB(5) graph shows that load voltage is a maximum of 26.99 dBV at the design frequency of 50 MHz, and falls off on either side. The 3 dB bandwidth would be the range of frequencies over which the magnitude is within 3 dB of 26.99 dBV. By estimation from the graph, this would be from 45 MHz to 55 MHz approximately. If greater accuracy is needed, the number of frequencies in the .AC control line could be increased or the starting (10 MHz) and stopping (90 MHz) frequencies could be changed.

CHAPTER SUMMARY

SPICE recognizes lossless transmission lines only. It is possible to add pseudo-distributed losses to a length of transmission line by breaking it up into many short lengths

```
********11/22/87******** Demo PSpice (May 1986) *******19:48:05********


MATCH.CIR   0.25 WAVELENGTH TRANS. LINE MATCH


****    CIRCUIT DESCRIPTION


**************************************************************************



VS    1  0  AC  141.42
RS    1  2  50
VSENSE 2 10  0
T1   10  0  3  0  Z0=50  F=50MEG  NL=2
T2    3  0  4  0  Z0=158.11  F=50MEG  NL=0.25
*     SQRT(500*50) = 158.11, 0.25 WAVE MATCH
T3    4  0  5  0  Z0=50  F=50MEG  NL=0.25
*     T3 SECTION TRANSFORMS 5 OHM RLOAD TO 500 OHM
RLOAD 5  0  5
.AC LIN 41  10MEG  90MEG
.PRINT AC  VM(10)  VP(10)   IM(VSENSE)  IP(VSENSE)  VM(5)
.PLOT AC VM(5) (5,25)
.PLOT AC VDB(5) (16,28)
.OPTIONS NOPAGE NODE
.END

****    ELEMENT NODE TABLE


      0    RLOAD    VS       T1       T1       T2       T2       T3
           T3

      1    RS       VS

      2    RS       VSENSE

      3    T1       T2

      4    T2       T3

      5    RLOAD    T3

     10    VSENSE   T1

****    SMALL SIGNAL BIAS SOLUTION      TEMPERATURE =   27.000 DEG C

  NODE   VOLTAGE      NODE   VOLTAGE      NODE   VOLTAGE      NODE   VOLTAGE


 ( 1)    .0000    ( 2)     .0000    ( 3)     .0000    ( 4)     .0000

 ( 5)    .0000    ( 10)    .0000
```

Figure 10.12 Output File MATCH.OUT.

```
****      AC ANALYSIS                       TEMPERATURE =   27.000 DEG C


    FREQ        VM(10)      VP(10)      IM(VSENSE)  IP(VSENSE)  VM(5)


1.000E+07    4.332E+01    6.398E+01    2.569E+00   -1.764E+01   9.006E+00
1.200E+07    1.087E+02    3.555E+01    1.649E+00   -5.005E+01   8.292E+00
1.400E+07    1.368E+02    3.678E+00    2.013E-01   -6.066E+01   7.722E+00
1.600E+07    1.205E+02   -2.794E+01    1.328E+00    5.827E+01   7.281E+00
1.800E+07    6.716E+01   -5.828E+01    2.410E+00    2.830E+01   6.953E+00
2.000E+07    6.982E+00    5.950E+01    2.760E+00   -2.498E+00   6.726E+00
2.200E+07    7.723E+01    5.414E+01    2.295E+00   -3.305E+01   6.591E+00
2.400E+07    1.255E+02    2.432E+01    1.167E+00   -6.233E+01   6.541E+00
2.600E+07    1.375E+02   -5.948E+00    2.998E-01    7.194E+01   6.577E+00
2.800E+07    1.104E+02   -3.598E+01    1.664E+00    5.121E+01   6.702E+00
3.000E+07    5.207E+01   -6.432E+01    2.556E+00    2.154E+01   6.925E+00
3.200E+07    2.128E+01    7.099E+01    2.720E+00   -8.510E+00   7.262E+00
3.400E+07    8.765E+01    4.808E+01    2.109E+00   -3.820E+01   7.739E+00
3.600E+07    1.287E+02    1.840E+01    8.993E-01   -6.461E+01   8.395E+00
3.800E+07    1.320E+02   -1.166E+01    5.862E-01    6.546E+01   9.289E+00
4.000E+07    9.581E+01   -4.063E+01    1.856E+00    4.224E+01   1.051E+01
4.200E+07    3.207E+01   -5.633E+01    2.530E+00    1.218E+01   1.219E+01
4.400E+07    4.605E+01    4.958E+01    2.339E+00   -1.745E+01   1.450E+01
4.600E+07    9.609E+01    2.512E+01    1.360E+00   -3.686E+01   1.752E+01
4.800E+07    9.674E+01   -2.250E+00    8.984E-01    4.850E+00   2.077E+01
5.000E+07    7.071E+01   -4.278E-13    1.414E+00    4.278E-13   2.236E+01
5.200E+07    9.674E+01    2.250E+00    8.984E-01   -4.850E+00   2.077E+01
5.400E+07    9.609E+01   -2.512E+01    1.360E+00    3.686E+01   1.752E+01
5.600E+07    4.605E+01   -4.958E+01    2.339E+00    1.745E+01   1.450E+01
5.800E+07    3.207E+01    5.633E+01    2.530E+00   -1.218E+01   1.219E+01
6.000E+07    9.581E+01    4.063E+01    1.856E+00   -4.224E+01   1.051E+01
6.200E+07    1.320E+02    1.166E+01    5.862E-01   -6.546E+01   9.289E+00
6.400E+07    1.287E+02   -1.840E+01    8.993E-01    6.461E+01   8.395E+00
6.600E+07    8.765E+01   -4.808E+01    2.109E+00    3.820E+01   7.739E+00
6.800E+07    2.128E+01   -7.099E+01    2.720E+00    8.510E+00   7.262E+00
7.000E+07    5.207E+01    6.432E+01    2.556E+00   -2.154E+01   6.925E+00
7.200E+07    1.104E+02    3.598E+01    1.664E+00   -5.121E+01   6.702E+00
7.400E+07    1.375E+02    5.948E+00    2.998E-01   -7.194E+01   6.577E+00
7.600E+07    1.255E+02   -2.432E+01    1.167E+00    6.233E+01   6.541E+00
7.800E+07    7.723E+01   -5.414E+01    2.295E+00    3.305E+01   6.591E+00
8.000E+07    6.982E+00   -5.950E+01    2.760E+00    2.498E+00   6.726E+00
8.200E+07    6.716E+01    5.828E+01    2.410E+00   -2.830E+01   6.953E+00
8.400E+07    1.205E+02    2.794E+01    1.328E+00   -5.827E+01   7.281E+00
8.600E+07    1.368E+02   -3.678E+00    2.013E-01    6.066E+01   7.722E+00
8.800E+07    1.087E+02   -3.555E+01    1.649E+00    5.005E+01   8.292E+00
9.000E+07    4.332E+01   -6.398E+01    2.569E+00    1.764E+01   9.006E+00
```

Figure 10.12 (Continued).

```
****    AC ANALYSIS                      TEMPERATURE =    27.000 DEG C


   FREQ      VM(5)

                  5.000D+00     7.477D+00    1.118D+01    1.672D+01    2.500D+01
                  - - - - - - - - - - - - - - - - - - - - - - - - - - - - - - -
 1.000D+07  9.006D+00 .              .        *     .            .            .
 1.200D+07  8.292D+00 .              .  *           .            .            .
 1.400D+07  7.722D+00 .              .*            .            .            .
 1.600D+07  7.281D+00 .           *. .            .            .            .
 1.800D+07  6.953D+00 .           * .            .            .            .
 2.000D+07  6.726D+00 .          * .            .            .            .
 2.200D+07  6.591D+00 .        *   .            .            .            .
 2.400D+07  6.541D+00 .        *   .            .            .            .
 2.600D+07  6.577D+00 .        *   .            .            .            .
 2.800D+07  6.702D+00 .        *   .            .            .            .
 3.000D+07  6.925D+00 .          * .            .            .            .
 3.200D+07  7.262D+00 .          *.             .            .            .
 3.400D+07  7.739D+00 .             .*           .            .            .
 3.600D+07  8.395D+00 .             .   *        .            .            .
 3.800D+07  9.299D+00 .             .      *     .            .            .
 4.000D+07  1.051D+01 .             .         *  .            .            .
 4.200D+07  1.219D+01 .             .            *.           .            .
 4.400D+07  1.450D+01 ..            .            .  *         .            .
 4.600D+07  1.752D+01 .             .            .         *  .            .
 4.800D+07  2.077D+01 .             .            .            .   *        .
 5.000D+07  2.236D+01 .             .            .            .      *     .
 5.200D+07  2.077D+01 .             .            .            .    *       .
 5.400D+07  1.752D+01 .             .            .         *  .            .
 5.600D+07  1.450D+01 .             .            .  *         .            .
 5.800D+07  1.219D+01 .             .            *.           .            .
 6.000D+07  1.051D+01 .             .         *  .            .            .
 6.200D+07  9.289D+00 .             .      *     .            .            .
 6.400D+07  8.395D+00 .             .   *        .            .            .
 6.600D+07  7.739D+00 .             .*           .            .            .
 6.800D+07  7.262D+00 .          *.             .            .            .
 7.000D+07  6.925D+00 .          * .            .            .            .
 7.200D+07  6.702D+00 .        *   .            .            .            .
 7.400D+07  6.577D+00 .        *   .            .            .            .
 7.600D+07  6.541D+00 .        *   .            .            .            .
 7.800D+07  6.591D+00 .        *   .            .            .            .
 8.000D+07  6.726D+00 .          * .            .            .            .
 8.200D+07  6.953D+00 .           *. .            .            .            .
 8.400D+07  7.281D+00 .           *. .            .            .            .
 8.600D+07  7.722D+00 .             .*           .            .            .
 8.800D+07  8.292D+00 .             .  *         .            .            .
 9.000D+07  9.006D+00 .             .        *   .            .            .
                  - - - - - - - - - - - - - - - - - - - - - - - - - - - - - - -
```

Figure 10.12 (Continued).

```
****    AC ANALYSIS                          TEMPERATURE =   27.000 DEG C


       FREQ      VDB(5)

                       1.600D+01      1.900D+01    2.200D+01    2.500D+01    2.800D+01
                     - - - - - - - - - - - - - - - - - - - - - - - - - - - - - - - - -
  1.000D+07  1.909D+01 .              *            .            .            .
  1.200D+07  1.837D+01 .            *   .          .            .            .
  1.400D+07  1.775D+01 .          *    .           .            .            .
  1.600D+07  1.724D+01 .       *       .           .            .            .
  1.800D+07  1.684D+01 .       *       .           .            .            .
  2.000D+07  1.656D+01 . *             .           .            .            .
  2.200D+07  1.638D+01 . *             .           .            .            .
  2.400D+07  1.631D+01 .*              .           .            .            .
  2.600D+07  1.636D+01 . *             .           .            .            .
  2.800D+07  1.652D+01 . *             .           .            .            .
  3.000D+07  1.681D+01 .      *        .           .            .            .
  3.200D+07  1.722D+01 .       *       .           .            .            .
  3.400D+07  1.777D+01 .          *    .           .            .            .
  3.600D+07  1.848D+01 .            *. .           .            .            .
  3.800D+07  1.936D+01 .              . *          .            .            .
  4.000D+07  2.043D+01 .              .    *       .            .            .
  4.200D+07  2.172D+01 .              .         *. .            .            .
  4.400D+07  2.323D+01 .              .           .    *        .            .
  4.600D+07  2.487D+01 .              .           .            *.            .
  4.800D+07  2.635D+01 .              .           .            .    *        .
  5.000D+07  2.699D+01 .              .           .            .            *.
  5.200D+07  2.635D+01 .              .           .            .    *        .
  5.400D+07  2.487D+01 .              .           .            *.            .
  5.600D+07  2.323D+01 .              .           .    *        .            .
  5.800D+07  2.172D+01 .              .         *. .            .            .
  6.000D+07  2.043D+01 .              .    *       .            .            .
  6.200D+07  1.936D+01 .              . *          .            .            .
  6.400D+07  1.848D+01 .            *. .           .            .            .
  6.600D+07  1.777D+01 .          *    .           .            .            .
  6.800D+07  1.722D+01 .       *       .           .            .            .
  7.000D+07  1.681D+01 .      *        .           .            .            .
  7.200D+07  1.652D+01 . *             .           .            .            .
  7.400D+07  1.636D+01 . *             .           .            .            .
  7.600D+07  1.631D+01 .*              .           .            .            .
  7.800D+07  1.638D+01 . *             .           .            .            .
  8.000D+07  1.656D+01 . *             .           .            .            .
  8.200D+07  1.684D+01 .       *       .           .            .            .
  8.400D+07  1.724D+01 .        *      .           .            .            .
  8.600D+07  1.775D+01 .          *    .           .            .            .
  8.800D+07  1.837D+01 .            *   .          .            .            .
  9.000D+07  1.909D+01 .              *            .            .            .
                     - - - - - - - - - - - - - - - - - - - - - - - - - - - - - - - - -


       JOB CONCLUDED

       TOTAL JOB TIME            17.00
```

Figure 10.12 (Continued).

with series resistance added. However, this technique may lead to substantially increased transient analysis computation time.

The element line for any transmission line must include the characteristic impedance, Z0, and an indication of the line length. The two ways of indicating length are transmission delay (in seconds) or normalized electrical length (in wavelengths) and frequency (in Hz).

Time-domain reflectometry problems (using transient analysis) and steady-state impedance and bandwidth solutions (using AC analysis) can be done easily with SPICE for transmission line circuits.

Chapter 11

HOW TO CHANGE SEMICONDUCTOR MODELS

11.1 WHY CHANGE SEMICONDUCTOR MODELS?

As you use SPICE to model semiconductor circuits, you will discover that the built-in semiconductor models for diode, BJT, JFET and MOSFET, with default parameters, do not always match the device characteristics of the semiconductors you are using. For example, you may have a circuit containing a superbeta bipolar transistor with a forward current gain, or beta, of 400. The default beta in a SPICE BJT is 100. Obviously the SPICE BJT model would have to be changed to simulate the circuit accurately. This is easily done.

Another instance in which you may want to change the parameters of a semiconductor model is when a circuit may be built with off-the-shelf components. Beta of BJTs and transconductance of FETs can vary widely from what manufacturers call "typical" values. Perhaps you designed the circuit assuming a typical beta of 100; the beta values of a production run of discrete transistors might vary from 40 to 250. In order to see how the modeled circuit performs with those extremes of beta, it is necessary for you to change the SPICE BJT model.

Educators may have occasion to use an "ideal" diode when teaching certain subjects such as power supply operation. The ideal diode would have essentially no forward voltage drop (behaves like a short circuit in forward conduction) and no reverse leakage current (acts like an open circuit in reverse bias). SPICE can closely approximate an ideal diode, but how to go about accomplishing this may not be obvious. A diode has 14 parameters that can be changed in the .MODEL control line, and the one you would think to change to make a diode ideal in forward conduction (VJ, or junction potential) turns out to have absolutely nothing to do with the diode forward voltage drop!

This chapter will provide an introduction to changing parameters for the four semiconductor devices SPICE recognizes. For the reader who wishes to dig deeper into this topic, there are some excellent references listed in the bibliography in Appendix D. Some of the suppliers of SPICE-based circuit analysis software include programs which can generate very accurate models for all types of semiconductors when given manufacturer's data sheet information or measurement data from the laboratory. These supplementary programs actually create the .MODEL line for the semiconductor, which can then be added to the circuit input file.

11.2 CHANGING THE SPICE DIODE MODEL

The table of diode model parameters in Chap. 8.2.3 lists 14 parameters. The DC characteristics are determined by IS (the saturation current) and N (the emission coefficient). Charge storage effects are specified by TT (transit time) and a nonlinear depletion layer capacitance which is determined by CJO (zero-bias junction capacitance), VJ (junction potential) and M (grading coefficient). The temperature dependence of the saturation current is determined by EG (activation energy) and XTI (saturation current temperature exponent). Reverse breakdown characteristics depend upon BV (reverse breakdown voltage) and IBV (current at breakdown voltage). Both BV and IBV should be expressed as positive numbers.

Yet another parameter which affects IS, RS, CJO and IBV is the area factor (AREA), which is listed not in the .MODEL line but rather in the DXXXXXX element line. The area factor determines how many of the diode model MODNAME are put in parallel to make one DXXXXXX.

11.2.1   The DC Diode Model

Figure 11.1 shows the DC model for a junction diode.   Note that the voltage VD is the junction voltage and does not include the ohmic voltage drop across the resistance RS.   The equation for diode current ID is

$$ID = IS \, [\exp(VD/N*VT) -1]$$

where IS is the saturation current, N is the emission coefficient (generally set to 1.0), and VT = k*T/q.   k is Boltzmann's constant, 1.38E-23 Joule/Kelvin, T is the absolute temperature in Kelvin, and q is the charge on an electron, 1.6E-19 C.   At a temperature of 25 deg C, k*T/q = VT = 26 mV.
     Solving the ID equation above for VD gives

$$VD = N * VT * \ln(ID/IS +1)$$

The two ways to change the forward voltage drop of the diode for a given forward current are to change N and to change IS.   Minor modification can be made to IS based on I-V measurements in the lab.   For example, the default value of IS is 1E-14 A.   This is a typical value for a silicon integrated circuit diode.   If you wanted to model a discrete germanium diode, a typical value of IS would be 5E-9 A.   It should be noted that IS also affects the reverse leakage current.

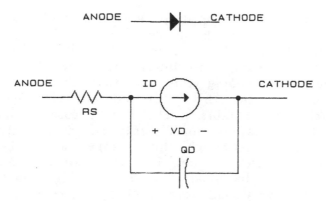

Figure 11.1   DC Model for a Diode.

EXAMPLE: How can the IS parameter be obtained?

Perhaps the simplest way is to obtain data which gives the diode voltage and current at one point. For example, if the forward voltage across a germanium diode was 316 mV when the diode current was 1 mA, the equation for ID above can be rearranged to give

$$IS = ID/[\exp(VD/NVT) -1]$$

Substituting the ID and VD values, and using a value of 1.0 for N and 26 mV for VT (the room temperature value), we obtain

$$IS = 1E-3/[\exp(0.316/(1.0*26E-3))-1] = 5E-9 = 5 \ nA$$

EXAMPLE: How can an ideal diode be approximated?

If one wanted to make a diode with a very low forward voltage drop, IS could be set to a very large value (such as 1E-3 A). However, by making this drastic change in IS an equally drastic change occurs in reverse leakage. Thus, by changing IS an ideal diode has been approximated in forward conduction only; such a diode would have unacceptable reverse leakage currents.

A table of forward voltage drops and reverse leakage currents appears below, for four kinds of diodes. The first is the SPICE default diode. The second and third diodes have had the IS parameter changed to lower the forward voltage drop; the consequences of doing this are that the reverse leakage current rises. In the third case, the reverse leakage current is clearly unacceptable. The fourth case shows a diode which has the emission coefficient, N, drastically reduced (from 1 to .001). Although this change cannot be justified based on semiconductor physics, it does produce a diode model that is quite close to ideal. It has a very small forward voltage drop and a very small reverse leakage current.

Model of Diode	IS	N	VD (at ID = 1 mA)	ID (at VD = -10 V)
SPICE default	1E-14	1	660 mV	-10 pA
IS changed	5E-9	1	316 mV	-5 nA
IS changed	1E-3	1	17.9 mV	-1 mA
N changed (almost ideal)	1E-14	.001	0.66 mV	-10 pA

The two circuits which were analyzed to generate the data above are shown in Figs. 11.2 and 11.3. Figure 11.2 shows a DC voltage source (VIN) across the parallel combination of four diodes (VA, VB, VC and VD are dead voltage sources used as ammeters). VIN is swept from 0 V to -10 V in -1 V steps. The current that flows through each diode is measured by the four dead voltage sources, and the results are presented in a table by the .PRINT statement.

The circuit of Fig. 11.3 shows a DC current source (IIN) connected to the four same diodes in series. IIN is swept from 0 to 20 mA in 1 mA steps, and the voltage across each diode is presented in a table by the .PRINT statement.

The SPICE input files DIODE-I.CIR and DIODE-V.CIR that were used to generate these data are shown in Figs. 11.4 and

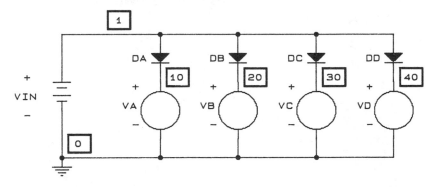

Figure 11.2 Circuit to Test Diode Leakage Current.

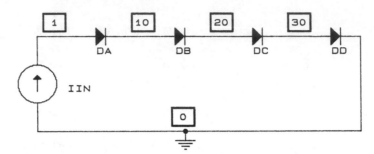

**Figure 11.3 Circuit to Test Diode Forward
Voltage.**

11.5 respectively; they may prove useful if you try modifying
diode parameters and want to see the results.

11.3   CHANGING SPICE TRANSISTOR MODELS

The table of bipolar transistor model parameters in Chap.
8.3.3 lists 40 parameters. Junction field-effect transistors
are characterized by 12 parameters, shown in Chap. 8.4.3.
Chapter 8.5.3 lists 42 parameters for MOSFETs. A meaningful
discussion of transistor parameters for BJT, JFET and MOSFET
transistors, and the methods used for creating accurate models of manufactured transistors, is beyond the scope of this

```
DIODE-I.CIR    TESTING EFFECT OF IS & N CHANGES ON REV DIODE CURRENT
*       THE SAME VOLTAGE IS ACROSS DIODES DA THROUGH DD
VIN 1 0 DC
DA 1 10 DA
DB 1 20 DB
DC 1 30 DC
DD 1 40 DD
VA 10 0
VB 20 0
VC 30 0
VD 40 0
.MODEL DA D
.MODEL DB D(IS=5E-9)
.MODEL DC D(IS=1E-3)
.MODEL DD D(N=.001)
.DC VIN 0 -10 -1
.PRINT DC I(VA) I(VB) I(VC) I(VD)
.OPTIONS NOPAGE
.OP
.END
```

Figure 11.4 Input File DIODE-I.CIR.

```
DIODE-V.CIR    TESTING EFFECT OF IS & N CHANGES ON DIODE VOLTAGE
*        THE SAME CURRENT FLOWS THROUGH DIODES DA THROUGH DD.
IIN 0 1 DC
DA 1 10 DA
DB 10 20 DB
DC 20 30 DC
DD 30 0 DD
.MODEL DA D
.MODEL DB D(IS=5E-9)
.MODEL DC D(IS=1E-3)
.MODEL DD D(N=.001)
.DC IIN 0 20M 1M
.PRINT DC V(1,10) V(10,20) V(20,30) V(30,0)
.OPTIONS NOPAGE
.OP
.END
```

Figure 11.5 Input File DIODE-V.CIR.

book.   There are some excellent references on this topic listed in Appendix D.

For many circuits, in order to simulate circuit behavior
accurately when a SPICE analysis is performed, the only transistor parameters that need to be changed are primary gain parameters.   These would be ideal maximum forward beta (BETA)
for a BJT, transconductance parameter (BETA) for a JFET, and
transconductance parameter (KP) for a MOSFET.   It is desirable to specify the minimum number of parameters for a semiconductor device which will provide a reasonably accurate model for two reasons:

1. Obtaining many of the parameters is often very difficult and time consuming; manufacturers' data sheets and
laboratory tests of transistors do not directly provide
the parameters which SPICE uses.   An example for a BJT is
the Early voltage, both forward and reverse.   These two
parameters (VAF and VAR) usually must be calculated from
other information (h parameters, or characteristic curves
which result from laboratory tests).

2. Computation time goes up as the number of parameters
specified for each semiconductor device increases.   Circuit designers generally work with simple models in order
to save time, and do achieve good results.   For the same
reason, the simplest model that will do the job should be
used when using SPICE.

Some circuits will have high-speed switching where the transition time is a function of the parameters of the transistor

(and to a lesser degree the parameters of the circuit). Examples are astable multivibrators and Schmitt trigger circuits. In such circuits it is necessary to include in the transistor model those parameters which will accurately reflect the device switching time. Failure to do this can result in the switching time being so small that SPICE will be forced into using computing time steps that are extremely small, and a transient analysis error may result.

11.3.1 Changing SPICE BJT Models

In order to obtain an accurate BJT model which gives realistic results in many circuits, the only parameter that needs to be changed is the ideal maximum forward beta, BF. It is the second parameter in the list of 40. The .MODEL line is used to set BF to other than its default value of 100. Typical values of BF range from 40 to 250. The element and model lines for three transistors with different forward betas would be

```
Q1  9 8 7 HIGH
Q2  6 5 4 MED
Q3  3 2 1 LOW
.MODEL HIGH  NPN(BF = 200)
.MODEL MED   NPN(BF = 120)
.MODEL LOW   NPN(BF = 40)
```

The switching time of a BJT is determined by its capacitances and its transit time. The capacitance parameters and their meanings are

```
CJE  B-E zero-bias depletion capacitance
CJC  B-C zero-bias depletion capacitance
CJS  zero-bias collector-substrate capacitance
```

CJS is used only for IC transistors, and would be zero for a discrete transistor. CJC and CJE may be available from data sheets. Typical values are 1 or 2 pF.

The other parameter which affects switching time is the ideal forward transit time, TF. This is seldom available from data sheets, but there is a simple way to calculate a good approximation to this parameter

$$TF = 1/(2*PI*fT)$$

where fT is called the common-emitter cutoff frequency, gain-bandwidth product or transition frequency. It is the frequency at which an extrapolation of the forward current gain on a graph of gain versus frequency falls to unity. Values of TF typically range from 0.1 ns to 20 ns for discrete transistors.

The default values of CJE, CJC and TF are zero. The element line and model line for a transistor with those parameters modified is

```
QOUTPUT  10  4  30   SPIFFY
.MODEL  SPIFFY  NPN(CJE = 1.2P  CJC = 0.8P  TF = 1.2N)
```

11.3.2 Changing SPICE JFET Models

In order to obtain an accurate JFET model which gives realistic results in many circuits, the only parameter that needs to be changed is the transconductance parameter, BETA. It is the second parameter in the list of 12. The .MODEL line is used to set BETA to other than its default value of 1.0E-4. The BETA parameter has units of amp/(volt*volt); this is different from the transconductance, gm, that is normally available from manufacturers' data sheets. The unit for gm is amp/volt. If the channel length modulation parameter, LAMBDA, is neglected (its default value is zero) the relationship between gate-source voltage, VGS, and drain current, ID, in the SPICE JFET model is given by

$$ID = BETA*(VGS - VTO)^2$$

where VTO is the threshold voltage of the JFET. Since the transconductance, gm, is the derivative of drain current with respect to gate-source voltage, transconductance is

$$gm = \frac{d(ID)}{d(VGS)} = 2(BETA)(VGS - VTO)$$

Re-arranging the equation above for BETA gives

$$BETA = gm/(2*(VGS - VTO).$$

Thus, if the data sheet for a JFET gives the parameters gm, VTO and VGS at a certain operating point the value of BETA can be calculated with the equation above. The element and model lines for three transistors with different transconductance parameters would be

```
J1  9 8 7  BIGGER
J2  6 5 4  SMALLER
.MODEL BIGGER    NJF(BETA = 5E-4)
.MODEL SMALLER PJF(BETA = 2E-4)
```

Notice that J1 is an N-channel JFET while J2 is a P-channel JFET.

11.3.3 Changing SPICE MOSFET Models

SPICE can simulate a MOSFET using three different models. The difference in the models is how the I-V characteristic is formulated. The MOSFET model has 42 parameters, of which 25 have default values of zero. The third parameter in the list of 42 is KP, the transconductance parameter. As with the JFET model, this transconductance can easily be changed to any desired value other than its default value of 2E-5 by using the .MODEL line. Shown below are four lines which describe two MOSFETs with different transconductance parameters.

```
M1  9 8 7 6  BIGGER
M2  5 4 3 2  SMALLER
.MODEL BIGGER  PMOS(KP = 3E-5)
.MODEL SMALLER  NMOS(KP = 1E-5)
```


CHAPTER SUMMARY

SPICE recognizes four types of semiconductors (diode, BJT, JFET and MOSFET) and has built-in models for each.

For many analyses, the built-in models need not be changed. If models need to be changed, generally only a few of the parameters have to be modified.

It is difficult to obtain most semiconductor parameters for the SPICE model from manufacturers' data sheets or laboratory measurements. Programs which generate the .MODEL line parameters from the above information are available from several suppliers of SPICE for personal computers.

An ideal diode (tiny voltage drop in forward conduction, very small leakage current in reverse bias) can be approximated by changing the emission coefficient, N, from the default value of 1.0 to 0.001. This is done in the diode .MODEL line.

Chapter 12

SAMPLE CIRCUITS

12.1 WHY SAMPLE CIRCUITS ?

Perhaps the best reference to use when writing SPICE input files is previous files of a similar type.   By looking back at successful SPICE analyses you can see how a particular type of analysis is specified, how to change temperature, how to specify a sub-circuit or how to print or plot a parameter of interest.   This rest of this chapter is divided into five sections, each of which has examples of specific kinds of analysis.

Section	Kinds of Examples
12.2	No analysis specified, SPICE does a small-signal bias solution anyhow.
12.3	DC analysis; one or more DC sources are varied and DC operating point is determined for each source value. Includes non-linear elements. Results may be printed and/or plotted.
12.4	AC analysis; a linear, small-signal analysis over a range of user-specified frequencies. Results may be printed and/or plotted.

152

12.5   <u>Transient analysis</u>; specified output variables
                are determined as functions of time.   Includes
                non-linear elements.   Results may be printed
                and/or plotted.

12.6   <u>.TEMP</u>; the circuit temperature may be set to
                any number of values (default temperature is 27
                degrees C).
                <u>.FOUR</u>; must be done in conjunction with a trans-
                ient analysis.   Performs a Fourier analysis of
                an output variable and gives amplitude and
                phase of first nine harmonics, as well as the
                DC component.
                <u>.TF</u>; does a small-signal transfer function anal-
                ysis of a circuit from a specified input to a
                specified output.   Results in input resistance,
                output resistance and transfer function (volt-
                age gain, current gain, transresistance or
                transconductance) being printed.
                <u>.OP</u>; SPICE will solve for the DC operating
                point of a circuit including many transistor pa-
                rameters, voltages and currents
                <u>.DISTO</u>; must be done in conjunction with an AC
                analysis.   Several kinds of distortion will be
                determined.   Results may be printed and/or
                plotted.
                <u>.SENS</u>; the small-signal sensitivities of one or
                more output variables, with respect to every pa-
                rameter in the circuit, will be found.
                <u>.NOISE</u>;   must be done in conjunction with an AC
                analysis.   SPICE will solve for the equivalent
                input and output noise of a circuit.   Results
                may be printed and/or plotted.

12.2 NO ANALYSIS SPECIFIED

EXAMPLE 12.2.1, DCMESS.CIR

Figure 12.1 shows a DC circuit with one battery and 9 resistors.   The problem is to find the DC voltage across RG, the
600 ohm resistor.   The SPICE input file is shown in Fig.
12.2.   It has element cards for each resistor and the battery, and no analysis is specified except for .OP.   This will
cause the voltage source current(s) to be printed in the output file.

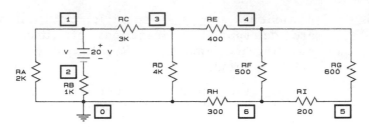

Figure 12.1 DC Circuit.

```
DCMESS.CIR    A CIRCUIT THAT COULD CAUSE A HEADACHE
V          1   2   20
RA         1   0   2K
RB         2   0   1K
RC         3   1   3K
RD         3   0   4K
RE         3   4   400
RF         4   6   500
RG         4   5   600
RH         6   0   300
RI         6   5   200
.OP
.OPTIONS NOPAGE
.END
```

Figure 12.2 Input File DCMESS.CIR.

The short output file in Fig. 12.3 contains the small-signal bias solution, which gives the voltage at each node compared to ground. Since we need to know the voltage across resistor RG, it will be necessary to manually subtract the voltage at node 5 from the voltage at node 4, as follows

$$V_{RG} = V(4) - V(5) = 1.4474 - 0.8977 = 0.5497 \text{ V}$$

EXAMPLE 12.2.2, VANDI.CIR

The DC circuit in Fig. 12.4 contains a voltage source, a current source and three resistors. The voltage across the 40 ohm resistor and the current through the 30 ohm resistor are to be found. Notice that a dead voltage source, V-AMP, is inserted in series with the 30 ohm resistor. It has no effect on the circuit operation and serves as an ammeter to give the current in that circuit branch.

```
********11/22/87******** Demo PSpice (May 1986) ********19:36:11********

DCMESS.CIR   A CIRCUIT THAT COULD CAUSE A HEADACHE

****    CIRCUIT DESCRIPTION

****************XX******************************************************


     V    1  2  20
     RA   1  0  2K
     RB   2  0  1K
     RC   3  1  3K
     RD   3  0  4K
     RE   3  4  400
     RF   4  6  500
     RG   4  5  600
     RH   6  0  300
     RI   6  5  200
     .OP
     .OPTIONS NOPAGE
     .END

****    SMALL SIGNAL BIAS SOLUTION     TEMPERATURE =   27.000 DEG C

  NODE   VOLTAGE     NODE   VOLTAGE     NODE   VOLTAGE     NODE   VOLTAGE


 ( 1)   11.3455   ( 2)   -8.6545   ( 3)    2.4001   ( 4)    1.4474

 ( 5)     .8977   ( 6)    .7145



     VOLTAGE SOURCE CURRENTS

     NAME       CURRENT

     V       -8.655D-03


     TOTAL POWER DISSIPATION   1.73D-01  WATTS


        JOB CONCLUDED

        TOTAL JOB TIME         3.80
```

Figure 12.3 Output File DCMESS.OUT

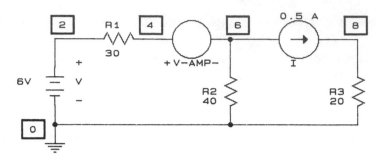

Figure 12.4 DC Circuit with "Ammeter".

Input file VANDI.CIR is shown in Fig. 12.5, and includes a .OP control line to cause an operating point analysis. The .OPTIONS . . . NODE control line causes an element node table to be printed; this lists each node and the elements that are connected to those nodes. Figure 12.6 shows the output file VANDI.OUT, which gives the node voltages. The voltage across the 40 ohm resistor is simply V(6), or -5.1429 V, since the bottom of that resistor is at node 0. The current through the 30 ohm resistor is the current through V-AMP, which is 3.714D-01 A, or 371.4 mA.

Notice that the current through DC source V is listed as -371.4 mA. The reason for the negative sign is that SPICE considers a current positive when it flows through a source from the + node to the - node. Since the current through DC source V is from node 0 to node 2, it is negative (source V is supplying power to the circuit).

```
VANDI.CIR   VOLTAGE SOURCE AND CURRENT SOURCE DC CIRCUIT
*       OBJECT IS TO SOLVE FOR VOLTAGE ACROSS 40 OHM R AND
*       CURRENT THROUGH 30 OHM R.
V     2  0  6
R1    2  4  30
*       THE LINE BELOW IS A DEAD VOLTAGE SOURCE = AMMETER
V-AMP 4  6  0
R2    6  0  40
I     6  8  500M
R3    8  0  20
.OP
.OPTIONS NOPAGE NODE
.END
```

Figure 12.5 Input File VANDI.CIR.

```
********11/22/87******** Demo PSpice (May 1986) ********20:00:52********

VANDI.CIR   VOLTAGE SOURCE AND CURRENT SOURCE DC CIRCUIT

****      CIRCUIT DESCRIPTION

*************************************************************************

*      OBJECT IS TO SOLVE FOR VOLTAGE ACROSS 10 OHM R AND
*      CURRENT THROUGH 30 OHM R.
V     2  0  6
R1    2  4  30
*      THE LINE BELOW IS A DEAD VOLTAGE SOURCE = AMMETER
V-AMP 4  6  0
R2    6  0  40
I     6  8  500M
R3    8  0  20
.OP
.OPTIONS NOPAGE NODE
.END

****      ELEMENT NODE TABLE

      0    R2      R3      V

      2    R1      V

      4    R1      V-AMP

      6    R2      V-AMP    I

      8    R3      I

****      SMALL SIGNAL BIAS SOLUTION      TEMPERATURE =   27.000 DEG C

   NODE   VOLTAGE      NODE   VOLTAGE      NODE   VOLTAGE      NODE   VOLTAGE

  ( 2)   6.0000    ( 4)   -5.1429    ( 6)   -5.1429    ( 8)   10.0000


      VOLTAGE SOURCE CURRENTS

      NAME        CURRENT

      V        -3.714D-01

      V-AMP     3.714D-01


      TOTAL POWER DISSIPATION   9.80D+00  WATTS

          JOB CONCLUDED

          TOTAL JOB TIME            4.30
```

Figure 12.6 Output File VANDI.OUT.

EXAMPLE 12.2.3, DC-CKT.CIR

The schematic diagram in Fig. 12.7 shows a DC circuit with a linear voltage-controlled current source, called G. The current through VCCS G is equal to five times VX, the voltage across the 15 ohm resistor. The problem is to find the voltage across the 6 ohm resistor and the current through the 25 ohm resistor. An ammeter (V-AMP) is put in series with the 25 ohm resistor for this purpose. In order to find the voltage across the 6 ohm resistor easily, ground (node 0) has been placed at its bottom.

The input file, Fig. 12.8, contains the control line .OP. This specifies an operating point analysis, which will cause the current through each source to be printed. The output file in Fig. 12.9 shows the voltage at node 3 to be 1.8907 V, and the current through V-AMP to be -318.2 mA.

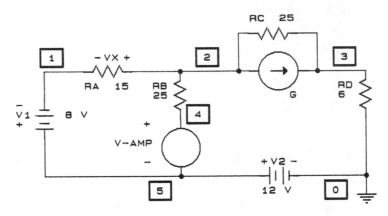

Figure 12.7   DC Circuit with Linear VCCS.

```
DC-CKT.CIR      CONTAINS A VOLTAGE-CONTROLLED CURRENT SOURCE
V1       5  1  8
RA       1  2  15
RB       2  4  25
*        THE LINE BELOW IS A DEAD VOLTAGE SOURCE = AMMETER
V-AMP    4  5  0
*        THE LINE BELOW IS A VOLTAGE-CONTROLLED CURRENT SOURCE
G        2  3  2  1  5
RC       2  3  25
RD       3  0  6
V2       5  0  12
.OP
.OPTIONS NOPAGE
.END
```

Figure 12.8   Input File DC-CKT.CIR.

```
********11/22/87******** Demo PSpice (May 1986) *******19:35:20********

DC-CKT.CIR    CONTAINS A VOLTAGE-CONTROLLED CURRENT SOURCE

****     CIRCUIT DESCRIPTION

***********************************************************************

V1      5  1  8
RA      1  2  15
RB      2  4  25
*       THE LINE BELOW IS A DEAD VOLTAGE SOURCE = AMMETER
V-AMP   4  5  0
*       THE LINE BELOW IS A VOLTAGE-CONTROLLED CURRENT SOURCE
G       2  3  2  1  5
RC      2  3  25
RD      3  0  6
V2      5  0  12
.OP
.OPTIONS NOPAGE
.END

****     SMALL SIGNAL BIAS SOLUTION      TEMPERATURE =   27.000 DEG C

  NODE   VOLTAGE     NODE   VOLTAGE     NODE   VOLTAGE     NODE   VOLTAGE

 (  1)    4.0000   (  2)    4.0458   (  3)    1.8907   (  4)   12.0000

 (  5)   12.0000


    VOLTAGE SOURCE CURRENTS

    NAME       CURRENT

    V1       -3.052D-03

    V-AMP    -3.182D-01

    V2       -3.151D-01

    TOTAL POWER DISSIPATION   3.81D+00  WATTS

****     OPERATING POINT INFORMATION     TEMPERATURE =   27.000 DEG C


**** VOLTAGE-CONTROLLED CURRENT SOURCES

             G
I-SOURCE  2.29E-01


         JOB CONCLUDED

         TOTAL JOB TIME          3.90
```

Figure 12.9 Output File DC-CKT.OUT.

Since the V-AMP current has a negative polarity, current is flowing upward through the 25 ohm resistor, from node 4 to node 2.

12.3 DC ANALYSIS

EXAMPLE 12.3.1, DI-VA.CIR

Figure 12.10 illustrates a test circuit which can generate a volt-ampere curve for a diode. The diode is the SPICE default diode, so the .MODEL line contains only the model name (PLAIN) and the type of semiconductor specification (D). All the default values for diode parameters are therefore used. The diode voltage is the same as the source (VD) voltage, and a dead voltage source (V-ID) has been put in series with the diode to measure current.

The input file is shown in Fig. 12.11. The diode voltage is stepped from 0.6 V to 0.81 V, in increments of 10 mV. A graph of diode current versus diode voltage is specified by the control line

.PLOT DC I(V-ID) (0, 0.4)

where the plot limits for diode current are 0 to 400 mA.

Figure 12.12 contains the output file DI-VA.OUT, which lists the IS parameter value (100 pA) and contains a graph of

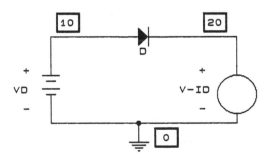

Figure 12.10 Diode Test Circuit.

```
DI-VA        DIODE V-A CHARACTERISTIC WITH .DC ANALYSIS
VD      10  0  DC
D       10  20 PLAIN
V-ID    20  0  0
.DC VD  0.6 0.81  .01
.MODEL PLAIN D
.PLOT  DC  I(V-ID)  (0, 0.4)
.OPTIONS NOPAGE
.END
```

Figure 12.11 Input File DI-VA.CIR.

```
********11/22/87******** Demo PSpice (May 1986) ********19:38:06********

DI-VA      DIODE V-A CHARACTERISTIC WITH .DC ANALYSIS

****     CIRCUIT DESCRIPTION

*************************************************************************

VD      10  0  DC
D       10  20 PLAIN
V-ID    20  0  0
.DC VD         0.6  0.81  .01
.MODEL PLAIN D
.PLOT  DC  I(V-ID)    (0, 0.4)
.OPTIONS NOPAGE
.END

****     DIODE MODEL PARAMETERS

           PLAIN

IS      1.00D-14

****     DC TRANSFER CURVES              TEMPERATURE =   27.000 DEG C

   VD        I(V-ID)

                  .000D+00    1.000D-01   2.000D-01   3.000D-01   4.000D-01
            - - - - - - - - - - - - - - - - - - - - - - - - - - - - - - - - -
 6.000D-01  1.188D-04 *            .           .           .           .
 6.100D-01  1.749D-04 *            .           .           .           .
 6.200D-01  2.574D-04 *            .           .           .           .
 6.300D-01  3.789D-04 *            .           .           .           .
 6.400D-01  5.578D-04 *            .           .           .           .
 6.500D-01  8.211D-04 *            .           .           .           .
 6.600D-01  1.209D-03 *            .           .           .           .
 6.700D-01  1.779D-03 *            .           .           .           .
 6.800D-01  2.619D-03 *            .           .           .           .
 6.900D-01  3.855D-03 .*           .           .           .           .
 7.000D-01  5.675D-03 .*           .           .           .           .
 7.100D-01  8.353D-03 .*           .           .           .           .
 7.200D-01  1.230D-02 . *          .           .           .           .
 7.300D-01  1.810D-02 . *          .           .           .           .
 7.400D-01  2.664D-02 .   *        .           .           .           .
 7.500D-01  3.922D-02 .      *     .           .           .           .
 7.600D-01  5.773D-02 .         *  .           .           .           .
 7.700D-01  8.498D-02 .           *.           .           .           .
 7.800D-01  1.251D-01 .            .  *        .           .           .
 7.900D-01  1.841D-01 .            .       * . .           .           .
 8.000D-01  2.711D-01 .            .           .       *   .           .
 8.100D-01  3.990D-01 .            .           .           .         *
            - - - - - - - - - - - - - - - - - - - - - - - - - - - - - - - - -

          JOB CONCLUDED
          TOTAL JOB TIME           5.40
```

Figure 12.12 Output File DI-VA.OUT.

current I(V-ID) versus diode voltage. At 600 mV, the diode current is 0.1188 mA, while at 810 mV the current has risen to 399.0 mA. Without the optional plot limits, SPICE would automatically have scaled the plot differently, resulting in a more compressed plot with less resolution.

EXAMPLE 12.3.2, FETINV.CIR

The junction field-effect transistor test circuit of Fig. 12.13 uses a SPICE default N-channel JFET. A graph of drain voltage versus gate-source voltage is desired. The gate-source voltage (VBATTERY) is stepped from -2.5 V to 0.5 V in steps of 100 mV.

Input file FETINV.CIR is shown in Fig. 12.14. The only specification for the JFET in the .MODEL line is NJF, so all the default parameters for JFETs will be used by SPICE. However, the element line for the JFET contains an area factor of 8. This means that the physical area of JFET J is 8 times larger than the default JFET, and several of the JFET parameters will be affected. See Chap. 8.4 for an explanation of the area factor. Notice that the element name of the JFET is simply J. If there were another JFET in the circuit, it would have to have another name, such as J2, JA, JBUFFER, etc.

Output file FETINV.OUT in Fig. 12.15 contains two of the JFET parameters (VTO and BETA) and a graph of drain voltage, V(3), versus gate-source voltage, V(4). Since V(3) was listed first in the .PLOT control line, it is plotted with asterisks and the values are printed as well as plotted. V(4)

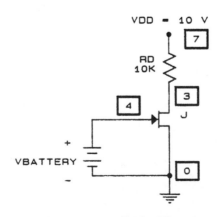

Figure 12.13 JFET Circuit.

```
FETINV.CIR   N-CHANNEL JFET INVERTER, TRANSFER CHARACTERISTIC
J      3  4  0  MOD1  8
.MODEL MOD1  NJF
VBATTERY 4 0 DC
VDD      7  0  10
RD       7  3  10K
.DC VBATTERY  -2.5  0.5  .1
.PLOT DC V(3) (0,10) V(4)
.OPTIONS NOPAGE
.END
```

Figure 12.14 Input File FETINV.CIR.

```
********11/22/87******** Demo PSpice (May 1986) *******19:43:35********


FETINV.CIR   N-CHANNEL JFET INVERTER, TRANSFER CHARACTERISTIC


****     CIRCUIT DESCRIPTION


*******************************************************************xxxxxxxxx


J      3  4  0  MOD1  8
.MODEL MOD1  NJF
VBATTERY 4 0 DC
VDD      7  0  10
RD       7  3  10K
.DC VBATTERY  -2.5  0.5  .1
.PLOT DC V(3) (0,10) V(4)
.OPTIONS NOPAGE
.END



****     JFET MODEL PARAMETERS



         MOD1

TYPE     NJF

VTO      -2.000

BETA     1.00D-04
```

Figure 12.15 Output File FETINV.OUT

```
****      DC TRANSFER CURVES                TEMPERATURE =    27.000 DEG C


LEGEND:

*: V(3)
+: V(4)

   VBATTERY     V(3)

(*)-------------    .000D+00    2.500D+00    5.000D+00    7.500D+00    1.000D+01
                    - - - - - - - - - - - - - - - - - - - - - - - - - - - - - -


(+)------------- -3.000D+00   -2.000D+00   -1.000D+00     .000D+00    1.000D+00
                                                - - - - - - - - - - - - - - - -
-2.500D+00  1.000D+01 .           +          .            .            .        *
-2.400D+00  1.000D+01 .            +         .            .            .        *
-2.300D+00  1.000D+01 .             +        .            .            .        *
-2.200D+00  1.000D+01 .              + .     .            .            .        *
-2.100D+00  1.000D+01 .               +.     .            .            .        *
-2.000D+00  1.000D+01 .               +      .            .            .        *
-1.900D+00  9.920D+00 .              .+      .            .            .       *
-1.800D+00  9.680D+00 .             . +      .            .            .    *  .
-1.700D+00  9.280D+00 .             .  +     .            .            .  *     .
-1.600D+00  8.720D+00 .             .   +    .            .            .*       .
-1.500D+00  8.000D+00 .             .     +  .            .          *          .
-1.400D+00  7.120D+00 .             .       +.            .        * .           .
-1.300D+00  6.080D+00 .             .          +          .      * .             .
-1.200D+00  4.880D+00 .             .            + *.     .                      .
-1.100D+00  3.520D+00 .             .          *         +.      .               .
-1.000D+00  2.000D+00 .             .     *    .           +      .              .
-9.000D-01  8.441D-01 .     *       .          .          .+      .              .
-8.000D-01  6.761D-01 .     *       .          .          .  +    .              .
-7.000D-01  5.838D-01 .   *         .          .          .   +   .              .
-6.000D-01  5.193D-01 .   *         .          .          .    +  .              .
-5.000D-01  4.709D-01 . *           .          .          .     + .              .
-4.000D-01  4.321D-01 . *           .          .          .      +.              .
-3.000D-01  4.000D-01 . *           .          .          .       +              .
-2.000D-01  3.729D-01 . *           .          .          .        +             .
-1.000D-01  3.496D-01 . *           .          .          .        +.            .
-1.110D-16  3.293D-01 . *           .          .          .         +            .
 1.000D-01  3.114D-01 . *           .          .          .          .+          .
 2.000D-01  2.955D-01 . *           .          .          .            +         .
 3.000D-01  2.813D-01 .*            .          .          .             +        .
 4.000D-01  2.684D-01 .*            .          .          .              +       .
 5.000D-01  2.568D-01 .*            .          .          .                +     .
                      - - - - - - - - - - - - - - - - - - - - - - - - - - - - - -



          JOB CONCLUDED

          TOTAL JOB TIME          7.60
```

Figure 12.15 (Continued).

is plotted with plus signs, and its value can only be estimated from the scale (-3 V to +1 V).

EXAMPLE 12.3.3, TTLINV.CIR

Figure 12.16 is a schematic diagram of a TTL 7404 inverter. The output is connected to a 2K ohm pull-up resistor. In order to determine the transfer characteristic of this circuit, a DC source (VIN) will be stepped from 1.3 V to 1.7 V in increments of 10 mV. This is the range of input voltage within which the output changes from logical 1 to logical 0 at room temperature.

The input file in Fig. 12.17 shows that the diode uses the SPICE model, and the four transistors use the SPICE model except that forward beta is set to 50 (default value is 100). The .PLOT control line requests a graph of output voltage and input voltage versus input voltage.

Contained in the output file of Fig. 12.18 are the diode saturation current, IS, and five of the transistor parameters (IS, BF, NF, BR AND NR). The graph shows that the output voltage is nearly 5 V until the input voltage increases to about 1.5 V. As the input voltage rises from 1.5 V to 1.62 V, the output monotonically decreases, although with a discontinuity at about 1.53 V. For input voltages above 1.63 V,

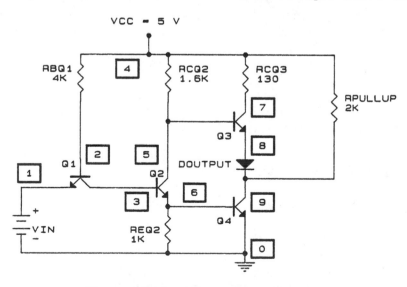

Figure 12.16 TTL Inverter Circuit.

```
TTLINV.CIR    7404 TTL INVERTER CIRCUIT, SWEPT DC INPUT
VCC      4    0   DC  5
VIN      1    0   DC
RBQ1     4    2   4K
RCQ2     4    5   1.6K
REQ2     6    0   1K
RCQ3     4    7   130
RPULLUP  4    9   2K
DOUTPUT  8    9   DI-MOD
Q1       3    2   1    Q-MOD
Q2       5    3   6    Q-MOD
Q3       7    5   8    Q-MOD
Q4       9    6   0    Q-MOD
.MODEL DI-MOD D
.MODEL Q-MOD NPN(BF = 50)
.OPTIONS NOPAGE
.DC  VIN  1.3  1.7  0.01
.PLOT DC  V(9,0)  (0,5)  V(1,0)  (1,2)
.END
```

Figure 12.17 Input File TTLINV.CIR.

the output is less than 30 mV, indicating that transistor Q4 is well into saturation.

One note of caution: in a circuit which is bi-stable, such as a Schmitt trigger (a TTL 7414 IC is one example), the use of DC analysis can be problematic. In the DC analysis, previous circuit values are not taken into consideration when calculating circuit conditions for the next input voltage. Since there are two possible stable output voltages for a given input voltage (depending on whether the input voltage is going up or going down), SPICE may fail to converge or may give incorrect results.

A better way to analyze a bi-stable circuit is to use a piece-wise linear (PWL) source with a triangular shape and perform a transient analysis. When a transient analysis is done SPICE does "remember" the previous circuit conditions when calculating at the next time increment.

12.4 AC ANALYSIS

EXAMPLE 12.4.1, LOPASS.CIR

The R-C circuit of Fig. 12.19 is connected to a 1 V sinusoid. It should have a half-power, or -3 dB, frequency of 500 Hz. SPICE AC analysis will be used to generate a Bode plot (graph of the magnitude and phase of the output voltage versus frequency) of the circuit response.

```
********11/22/87******** Demo PSpice (May 1986) *******19:59:29********

TTLINV.CIR   7404 TTL INVERTER CIRCUIT, SWEPT DC INPUT

****    CIRCUIT DESCRIPTION


**********************************************************************

VCC     4  0  DC 5
VIN     1  0  DC
RBQ1    4  2  4K
RCQ2    4  5  1.6K
REQ2    6  0  1K
RCQ3    4  7  130
RPULLUP       4  9  2K
DOUTPUT       8  9  DI-MOD
Q1      3  2  1  Q-MOD
Q2      5  3  6  Q-MOD
Q3      7  5  8  Q-MOD
Q4      9  6  0  Q-MOD
.MODEL DI-MOD D
.MODEL Q-MOD NPN(BF = 50)
.OPTIONS NOPAGE
.DC VIN 1.3 1.7 0.01
.PLOT DC  V(9,0)  (0,5) V(1,0) (1,2)
.END

****    DIODE MODEL PARAMETERS


            DI-MOD

IS      1.00D-14

****    BJT MODEL PARAMETERS


            Q-MOD

TYPE     NPN

IS      1.00D-16

BF        50.000

NF         1.000

BR         1.000

NR         1.000
```

Figure 12.18 Output File TTLINV.OUT.

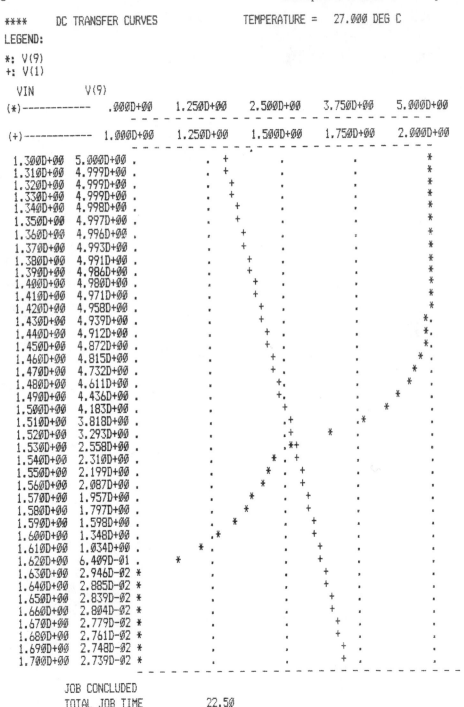

```
****    DC TRANSFER CURVES              TEMPERATURE =   27.000 DEG C
LEGEND:

*: V(9)
+: V(1)

  VIN         V(9)
(*)------------  .000D+00   1.250D+00   2.500D+00   3.750D+00   5.000D+00
                - - - - - - - - - - - - - - - - - - - - - - - - - - - - -
(+)------------  1.000D+00   1.250D+00   1.500D+00   1.750D+00   2.000D+00
                - - - - - - - - - - - - - - - - - - - - - - - - - - - - -
 1.300D+00  5.000D+00 .          . +         .           .           *
 1.310D+00  4.999D+00 .          . +         .           .           *
 1.320D+00  4.999D+00 .          .  +        .           .           *
 1.330D+00  4.999D+00 .          .  +        .           .           *
 1.340D+00  4.998D+00 .          .   +       .           .           *
 1.350D+00  4.997D+00 .          .   +       .           .           *
 1.360D+00  4.996D+00 .          .    +      .           .           *
 1.370D+00  4.993D+00 .          .    +      .           .           *
 1.380D+00  4.991D+00 .          .     +     .           .           *
 1.390D+00  4.986D+00 .          .     +     .           .           *
 1.400D+00  4.980D+00 .          .      +    .           .           *
 1.410D+00  4.971D+00 .          .      +    .           .           *
 1.420D+00  4.958D+00 .          .       +   .           .           *
 1.430D+00  4.939D+00 .          .       +   .           .         *.
 1.440D+00  4.912D+00 .          .        +  .           .         *.
 1.450D+00  4.872D+00 .          .        +  .           .         *.
 1.460D+00  4.815D+00 .          .         + .           .        * .
 1.470D+00  4.732D+00 .          .         + .           .       *  .
 1.480D+00  4.611D+00 .          .          +.           .      *   .
 1.490D+00  4.436D+00 .          .          +.           .     *    .
 1.500D+00  4.183D+00 .          .           +           .   *      .
 1.510D+00  3.818D+00 .          .           .+          .  *       .
 1.520D+00  3.293D+00 .          .           .+          *          .
 1.530D+00  2.558D+00 .          .           .*+         .          .
 1.540D+00  2.310D+00 .          .          * . +        .          .
 1.550D+00  2.199D+00 .          .        * . +          .          .
 1.560D+00  2.087D+00 .          .        * .  +         .          .
 1.570D+00  1.957D+00 .          .       * .    +        .          .
 1.580D+00  1.797D+00 .          .     * .      +        .          .
 1.590D+00  1.598D+00 .          .   * .        +        .          .
 1.600D+00  1.348D+00 .          . * .          +        .          .
 1.610D+00  1.034D+00 .        * .  .           +        .          .
 1.620D+00  6.409D-01 .     *    .  .           .        +          .
 1.630D+00  2.946D-02 *         .  .           .          +         .
 1.640D+00  2.885D-02 *         .  .           .          +         .
 1.650D+00  2.839D-02 *         .  .           .           +        .
 1.660D+00  2.804D-02 *         .  .           .           +        .
 1.670D+00  2.779D-02 *         .  .           .            +       .
 1.680D+00  2.761D-02 *         .  .           .            +       .
 1.690D+00  2.748D-02 *         .  .           .             +      .
 1.700D+00  2.739D-02 *         .  .           .             +      .
                - - - - - - - - - - - - - - - - - - - - - - - - - - - - -
          JOB CONCLUDED
          TOTAL JOB TIME        22.50
```

Figure 12.18 (Continued).

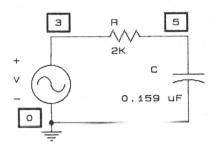

Figure 12.19   R-C Low-Pass Filter Circuit.

The input file LOPASS.CIR in Fig. 12.20 calls for the source voltage V to be stepped from 5 Hz to 0.5 MHz logarithmically, with 5 steps per decade of frequency. The output voltage (at node 5) is to be plotted in decibels.

Figure 12.21, LOPASS.OUT, contains a very uninteresting small signal bias solution (everything is zero) since there are no DC sources in the circuit. The graph of VDB(5) shows that the output voltage magnitude is essentially 0 dB at 5 Hz, falls to -3.006 dB at 500 Hz, and then linearly (on a logarithmic frequency axis) falls off at -20 dB/decade. For example, 500 KHz is 3 decades of frequency above 500 Hz, and the output at 500 KHz is -59.99 dB.

The phase of the output voltage starts at 0 degrees, is -45 degrees at 500 Hz, and asymptotically approaches -90 degrees as frequency increases above 500 Hz.

EXAMPLE 12.4.2, WEIN.CIR

The Wein bridge network is frequently used to determine the frequency of a low frequency R-C oscillator. Figure 12.22 shows half of a Wein bridge, which is the interesting and more difficult to understand half (the other half is two resistors in series). In order to see the response of the network with the output at node 6, the input voltage will be

```
LOPASS.CIR   1-POLE R-C LOW PASS FILTER, AC ANALYSIS
V        3  0  AC  1
R        3  5  2K
C        5  0  .159U
*        THE RC VALUES GIVE A CORNER FREQUENCY OF 500 HZ
.AC  DEC  5  5  0.5MEG
.PLOT  AC  VDB(5)  (-60,0)  VP(5)  (-90,0)
.OPTIONS  NOPAGE
.END
```

Figure 12.20   Input File LOPASS.CIR.

```
********11/22/87******** Demo PSpice (May 1986) *******19:47:19*********

LOPASS.CIR   1-POLE R-C LOW PASS FILTER, AC ANALYSIS

****     CIRCUIT DESCRIPTION

**************************************************************************
V      3   0   AC  1
R      3   5   2K
C      5   0   .159U
*        THE RC VALUES GIVE A CORNER FREQUENCY OF 500 HZ
.AC  DEC  5  5  0.5MEG
.PLOT  AC  VDB(5)  (-60,0)  VP(5)  (-90,0)
.OPTIONS  NOPAGE
.END

****     SMALL SIGNAL BIAS SOLUTION       TEMPERATURE =   27.000 DEG C

  NODE    VOLTAGE      NODE    VOLTAGE

 (  3)     .0000    (  5)      .0000


****     AC ANALYSIS                      TEMPERATURE =   27.000 DEG C
LEGEND:
*: VDB(5)
+: VP(5)
     FREQ       VDB(5)

(*)------------- -6.000D+01   -4.500D+01   -3.000D+01   -1.500D+01    .000D+00
                 - - - - - - - - - - - - - - - - - - - - - - - - - - -
(+)------------- -9.000D+01   -6.750D+01   -4.500D+01   -2.250D+01    .000D+00
                 - - - - - - - - - - - - - - - - - - - - - - - - - - -
  5.000D+00 -4.333D-04 .                  .             .             .           X
  7.924D+00 -1.089D-03 .                  .             .             .          +*
  1.256D+01 -2.734D-03 .                  .             .             .          +*
  1.991D+01 -6.864D-03 .                  .             .             .          +*
  3.155D+01 -1.722D-02 .                  .             .             .        + *
  5.000D+01 -4.313D-02 .                  .             .             .      +   *
  7.924D+01 -1.075D-01 .                  .             .             .    +     *
  1.256D+02 -2.652D-01 .                  .             .             .  +       *
  1.991D+02 -6.378D-01 .                  .             .             .+        *.
  3.155D+02 -1.453D+00 .                  .             .          +  .         *.
  5.000D+02 -3.006D+00 .                  .             .        +    .        *  .
  7.924D+02 -5.449D+00 .                  .             .  +          .      *    .
  1.256D+03 -8.632D+00 .                  .           +             .    *      .
  1.991D+03 -1.226D+01 .             +   .             .           . *
  3.155D+03 -1.610D+01 .       +         .             .          *.          .
  5.000D+03 -2.003D+01 .  +              .             .        *   .          .
  7.924D+03 -2.401D+01 . +               .             .      *     .          .
  1.256D+04 -2.800D+01 .+                .             .    *       .          .
  1.991D+04 -3.199D+01 .+                .             .  *  .               .
  3.155D+04 -3.599D+01 .+                .             *    .               .
  5.000D+04 -3.999D+01 +                 .          *       .               .
  7.924D+04 -4.399D+01 +                 .       .*         .               .
  1.256D+05 -4.799D+01 +                 .    *  .          .               .
  1.991D+05 -5.199D+01 +            *    .          .               .
  3.155D+05 -5.599D+01 +    *            .             .          .               .
  5.000D+05 -5.999D+01 X                 .             .          .               .
                       - - - - - - - - - - - - - - - - - - - - - - - - - - -

          JOB CONCLUDED
          TOTAL JOB TIME         5.70
```

Figure 12.21 Output File LOPASS.OUT.

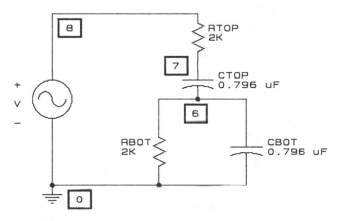

Figure 12.22 Wein Circuit.

swept logarithmically from 10 Hz to 1000 Hz with 15 frequencies per decade.

The input file in Fig. 12.23 calls for the output voltage magnitude and phase to be printed in a table and plotted on a graph versus frequency.   Notice that the two resistors are the same value, and the two capacitors have the same value, although they are specified differently.

As shown in Fig. 12.24, the table of output voltage in polar form does not lend itself readily to interpretation. One can easily note, however, that the phase becomes nearly zero (0.01081 degrees) at 100 Hz , and is positive below and negative above that frequency.   The magnitude and phase of the output are easily seen in the graph of VM(6) and VP(6) versus frequency.   The Wein circuit produces zero degrees of phase shift at only one frequency, which is the frequency where the output amplitude is a maximum.

Since at 100 Hz the input voltage is 1 V, 0 degrees, and the output voltage is 0.3333 V, 0 degrees, the gain of the circuit is 1/3, 0 degrees.   In order to make an oscillator with this circuit, a non-inverting, or angle 0 degrees gain

```
WEIN.CIR    WEIN-BRIDGE NETWORK, EQUAL R, EQUAL C
V        8   0   AC   1
RTOP     8   7   2K
CTOP     7   6   796N
RBOT     6   0   2E3
CBOT     6   0   .796U
.AC  DEC  15  10  1000
.PRINT  AC   VM(6)   VP(6)
.PLOT   AC   VM(6)   VP(6)
.OPTIONS NOPAGE
.END
```

Figure 12.23 Input File WEIN.CIR.

```
********11/22/87******** Demo PSpice (May 1986) *******20:02:40********

WEIN.CIR    WEIN-BRIDGE NETWORK, EQUAL R, EQUAL C

****    CIRCUIT DESCRIPTION

*****************************************************************************
V       8   0   AC  1
RTOP    8   7   2K
CTOP    7   6   796N
RBOT    6   0   2E3
CBOT    6   0   .796U
.AC DEC 15  10  1000
.PRINT  AC  VM(6)   VP(6)
.PLOT   AC  VM(6)   VP(6)
.OPTIONS NOPAGE
.END
****    SMALL SIGNAL BIAS SOLUTION        TEMPERATURE =   27.000 DEG C

  NODE   VOLTAGE       NODE   VOLTAGE     NODE   VOLTAGE

 ( 6)     .0000      ( 7)      .0000    ( 8)      .0000

****    AC ANALYSIS                      TEMPERATURE =   27.000 DEG C

     FREQ      VM(6)       VP(6)
   1.000E+01   9.669E-02   7.314E+01
   1.166E+01   1.114E-01   7.047E+01
   1.359E+01   1.279E-01   6.743E+01
   1.585E+01   1.462E-01   6.399E+01
   1.848E+01   1.660E-01   6.014E+01
   2.154E+01   1.871E-01   5.586E+01
   2.512E+01   2.090E-01   5.118E+01
   2.929E+01   2.310E-01   4.613E+01
   3.415E+01   2.525E-01   4.076E+01
   3.981E+01   2.725E-01   3.516E+01
   4.642E+01   2.904E-01   2.939E+01
   5.412E+01   3.056E-01   2.353E+01
   6.310E+01   3.177E-01   1.763E+01
   7.356E+01   3.264E-01   1.173E+01
   8.577E+01   3.316E-01   5.855E+00
   1.000E+02   3.333E-01  -1.081E-02
   1.166E+02   3.316E-01  -5.877E+00
   1.359E+02   3.263E-01  -1.176E+01
   1.585E+02   3.176E-01  -1.765E+01
   1.848E+02   3.056E-01  -2.355E+01
   2.154E+02   2.904E-01  -2.941E+01
   2.512E+02   2.725E-01  -3.518E+01
   2.929E+02   2.524E-01  -4.078E+01
   3.415E+02   2.309E-01  -4.615E+01
   3.981E+02   2.089E-01  -5.120E+01
   4.642E+02   1.870E-01  -5.588E+01
   5.412E+02   1.659E-01  -6.015E+01
   6.310E+02   1.461E-01  -6.401E+01
   7.356E+02   1.279E-01  -6.744E+01
   8.577E+02   1.114E-01  -7.048E+01
   1.000E+03   9.664E-02  -7.315E+01
```

Figure 12.24 Output File WEIN.OUT

```
****    AC ANALYSIS                        TEMPERATURE =   27.000 DEG C


LEGEND:

*: VM(6)
+: VP(6)

    FREQ      VM(6)

(*)------------ 6.310D-02   1.000D-01   1.585D-01   2.512D-01   3.981D-01
                - - - - - - - - - - - - - - - - - - - - - - - - - - - -

(+)------------ -1.000D+02  -5.000D+01   .000D+00   5.000D+01   1.000D+02
                - - - - - - - - - - - - - - - - - - - - - - - - - - - -

 1.000D+01  9.669D-02 .              *.          .            .       +    .
 1.166D+01  1.114D-01 .                . *        .            .      +     .
 1.359D+01  1.279D-01 .                  .   *    .            .      +     .
 1.585D+01  1.462D-01 .                  .      * .            .      +     .
 1.840D+01  1.660D-01 .                  .       .*            .     +      .
 2.154D+01  1.871D-01 .                  .       .    *        .   . +      .
 2.512D+01  2.090D-01 .                  .       .        *    . +          .
 2.929D+01  2.310D-01 .                  .       .            *+.           .
 3.415D+01  2.525D-01 .                  .       .            + *           .
 3.981D+01  2.725D-01 .                  .       .          +  . *          .
 4.642D+01  2.904D-01 .                  .       .        +     . *         .
 5.412D+01  3.056D-01 .                  .       .      +       .   *       .
 6.310D+01  3.177D-01 .                  .       .    +         .    *      .
 7.356D+01  3.264D-01 .                  .       .  +           .    *      .
 8.577D+01  3.316D-01 .                  .       .+             .     *     .
 1.000D+02  3.333D-01 .                  .      +.              .     *     .
 1.166D+02  3.316D-01 .                  .    + .               .     *     .
 1.359D+02  3.263D-01 .                  .   +  .               .    *      .
 1.585D+02  3.176D-01 .                  . +    .               .    *      .
 1.848D+02  3.056D-01 .                  +      .               .   *       .
 2.154D+02  2.904D-01 .                . +      .               .  *        .
 2.512D+02  2.725D-01 .               +  .      .               . *         .
 2.929D+02  2.524D-01 .              +   .      .               *           .
 3.415D+02  2.309D-01 .            .+    .      .            *  .            .
 3.981D+02  2.089D-01 .            +     .      .         *     .            .
 4.642D+02  1.870D-01 .          +  .    .      .       *       .            .
 5.412D+02  1.659D-01 .        +    .    .      . *             .            .
 6.310D+02  1.461D-01 .       +     .    .    * .               .            .
 7.356D+02  1.279D-01 .     +       .    *     .               .            .
 8.577D+02  1.114D-01 .     +    . *      .               .            .
 1.000D+03  9.664D-02 .    +   *.         .               .            .
                - - - - - - - - - - - - - - - - - - - - - - - - - - - -


        JOB CONCLUDED

        TOTAL JOB TIME        7.70
```

Figure 12.24 (Continued).

of 1/(1/3), or 3 would have to be provided. This is true only if the resistors are equal and the capacitors are equal.

EXAMPLE 12.4.3, LCMATCH.CIR

The circuit in Fig. 12.25 would be troublesome at best to
analyze by hand at a single frequency. It is a matching network, which is to be used at several frequencies. However,
ROUT and COUT, the resistive and reactive parts of the load
impedance, vary with frequency. For this reason the analysis
will be at 2 MHz only.

    The input file of Fig. 12.26 calls for an AC analysis at
one frequency, with six voltages to be printed. The load resistance is ROUT between nodes 5 and 6, and the voltage
across ROUT is to be printed as a magnitude and in decibel
form.

    If the 2 V input source (VIN and RIN) were connected to a
matched resistive load of 50 ohms, the load would have 1 V
across it (and RIN would drop 1 V). Figure 12.27 is the out-

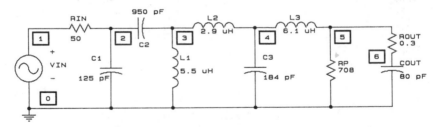

Figure 12.25 LCR Matching Network.

```
LCMATCH.CIR      L-C MATCHING NETWORK, SINGLE FREQ. AC ANALYSIS
VIN      1   0   AC  2
RIN      1   2   50
C1       2   0   125P
C2       2   3   950P
L1       3   0   5.5U
L2       3   4   2.9U
C3       4   0   184P
L3       4   5   6.1U
RP       5   0   710
ROUT     5   6   0.3
COUT     6   0   80P
*        CONTROL LINES FOLLOW
.AC LIN  1   2MEG   2MEG
.PRINT AC  VM(2)  VM(2,3)  VM(4)   VM(6)
.PRINT AC  VM(5,6)  VDB(5,6)
.OPTIONS NOPAGE
.END
```

Figure 12.26 Input File LCMATCH.CIR.

```
********11/22/87******** Demo PSpice (May 1986) ********19:45:24********

LCMATCH.CIR    L-C MATCHING NETWORK, SINGLE FREQ. AC ANALYSIS

****      CIRCUIT DESCRIPTION

*************************************************************************
VIN    1  0  AC  2
RIN    1  2  50
C1     2  0  125P
C2     2  3  950P
L1     3  0  5.5U
L2     3  4  2.9U
C3     4  0  184P
L3     4  5  6.1U
RP     5  0  710
ROUT   5  6  0.3
COUT   6  0  80P
*      CONTROL LINES FOLLOW
.AC  LIN  1  2MEG  2MEG
.PRINT AC  VM(2)  VM(2,3)  VM(4)  VM(6)
.PRINT AC  VM(5,6)  VDB(5,6)
.OPTIONS NOPAGE
.END

****      SMALL SIGNAL BIAS SOLUTION      TEMPERATURE =   27.000 DEG C

 NODE    VOLTAGE      NODE    VOLTAGE      NODE    VOLTAGE      NODE    VOLTAGE


(  1)    .0000     (  2)     .0000     (  3)     .0000     (  4)     .0000

(  5)    .0000     (  6)     .0000

****      AC ANALYSIS                     TEMPERATURE =   27.000 DEG C


    FREQ        VM(2)        VM(2,3)      VM(4)       VM(6)


  2.000E+06    5.171E-01    2.516E+00    3.038E+00   3.270E+00

****      AC ANALYSIS                     TEMPERATURE =   27.000 DEG C


    FREQ        VM(5,6)      VDB(5,6)


  2.000E+06    9.861E-04   -6.012E+01

          JOB CONCLUDED

          TOTAL JOB TIME            5.10
```

Figure 12.27 Output File LCMATCH.OUT.

put file which shows that the voltage across the load ROUT is 0.9861 mV, or -60.12 dB below 1 V.   Obviously, this is not a high-efficiency matching network.   Some of the voltages across reactances (VM(4) and VM(6), for example) exceed the input voltage; this is indicative of some resonance effects occurring in the matching network.

One of the limitations of SPICE is shown here, which is that a load which is frequency-dependent can not easily be described to SPICE.   Antenna impedances are one common example of this type of load.   Since the load impedance is a function of frequency, the input file must be edited (for load impedance) and a SPICE analysis must be done separately at each frequency.

EXAMPLE 12.4.4, SERIESAC.CIR

It is sometimes useful to be able to graphically compare the performance of two circuits.   In this example the frequency response of two series resonant circuits which are identical except for the loss resistance in each will be compared. SPICE will plot the results on one frequency axis, to make the comparison easy.

Figure 12.28 illustrates the two circuits which are connected in parallel across a voltage source; in this way a single plot of two response curves can be made.   A similar technique can be applied when a current source is the input by placing the two circuits in series.

Each circuit is resonant at 20 KHz; they differ in the series loss resistance, which determines quality factor Q and

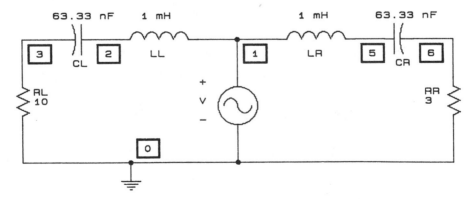

Figure 12.28 Two Series Resonant Circuits.

```
SERIESAC.CIR   TWO SERIES RESONANT CIRCUITS, ONE SOURCE
V         1  0  AC  1
*         LEFT SIDE SERIES CKT FOLLOWS, Q = 13
LL        1  2  1M
CL        2  3  63.33N
RL        3  0  10
*         RIGHT SIDE SERIES CKT FOLLOWS, Q = 42
LR        1  5  1E-3
CR        5  6  6.33E-8
RR        6  0  3
*         CONTROL LINES FOLLOW
.AC  LIN  21  15K  25K
.PLOT  AC  V(3)  V(6)  (.1,1)
.OPTIONS NOPAGE
.END
```

Figure 12.29   Input File SERIESAC.CIR.

bandwidth.   The source, V, will be swept linearly from 15 KHz
to 25 KHz, and the voltage across the two load resistors will
be plotted.   The input file, Fig. 12.29, shows the control
lines which will accomplish this.

The graph in the output file of Fig. 12.30 shows the
graph of resistor voltage versus frequency for the two cir-

```
********11/22/87******** Demo PSpice (May 1986) *******19:53:18********


SERIESAC.CIR   TWO SERIES RESONANT CIRCUITS, ONE SOURCE


****     CIRCUIT DESCRIPTION


*************************************************************************



V         1  0  AC  1
*         LEFT SIDE SERIES CKT FOLLOWS, Q = 13
LL        1  2  1M
CL        2  3  63.33N
RL        3  0  10
*         RIGHT SIDE SERIES CKT FOLLOWS, Q = 42
LR        1  5  1E-3
CR        5  6  6.33E-8
RR        6  0  3
*         CONTROL LINES FOLLOW
.AC  LIN  21  15K  25K
.PLOT  AC  V(3)  V(6)  (.1,1)
.OPTIONS NOPAGE
.END
```

Figure 12.30   Output File SERIESAC.OUT.

```
****     SMALL SIGNAL BIAS SOLUTION        TEMPERATURE =   27.000 DEG C

  NODE   VOLTAGE      NODE   VOLTAGE      NODE   VOLTAGE      NODE   VOLTAGE


(  1)    .0000    (  2)     .0000    (  3)      .0000    (  5)      .0000

(  6)    .0000

****     AC ANALYSIS                       TEMPERATURE =   27.000 DEG C


LEGEND:

*: V(3)
+: V(6)

    FREQ      V(3)

(*+)------------  1.000D-01    1.778D-01    3.162D-01    5.623D-01    1.000D+00
                  - - - - - - - - - - - - - - - - - - - - - - - - - - - - - -
 1.500D+04  1.352D-01 .        *         .            .            .            .
 1.550D+04  1.526D-01 .         *        .            .            .            .
 1.600D+04  1.742D-01 .           *      .            .            .            .
 1.650D+04  2.014D-01 .           .   *  .            .            .            .
 1.700D+04  2.369D-01 .           .      *  .         .            .            .
 1.750D+04  2.849D-01 .           .      .      *  .  .            .            .
 1.800D+04  3.528D-01 .  +        .      .         .* .            .            .
 1.850D+04  4.544D-01 .           .  +   .         .  .      *     .            .
 1.900D+04  6.130D-01 .           .      .  +      .  .            .*           .
 1.950D+04  8.440D-01 .           .      .         .  .   +        .    *       .
 2.000D+04  1.000D+00 .           .      .         .  .            .   +        . *            X
 2.050D+04  8.493D-01 .           .      .         .  .            .   +        .    *
 2.100D+04  6.316D-01 .           .      .  +      .  .            .       *    .
 2.150D+04  4.816D-01 .           .  +   .         .  .      *     .
 2.200D+04  3.846D-01 .        +  .      .         .  .   *        .
 2.250D+04  3.193D-01 +         .        .         .* .            .
 2.300D+04  2.729D-01 .           .      .      *  .  .            .
 2.350D+04  2.385D-01 .           .      .   *     .  .            .
 2.400D+04  2.121D-01 .           .      . *       .  .            .
 2.450D+04  1.911D-01 .           .   .* .         .  .            .
 2.500D+04  1.741D-01 .           *      .         .  .            .
                  - - - - - - - - - - - - - - - - - - - - - - - - - - - - - -


    JOB CONCLUDED

    TOTAL JOB TIME           6.40
```

Figure 12.30 (Continued).

cuits.   SPICE normally would scale each plot differently; in order to compare the two responses, the plot limit (0.1,1) was put in the .PLOT control line.   This plot limit affects both V(3) and V(6).   Because of this, some of the data points for V(6) do not fit on the graph.

However, it is easy to see that at 20 KHz both circuits have the same maximum output, indicated by an X on the graph.   The V(3) plot (asterisks) shows a much wider bandwidth than the V(6) plot (plus signs).   This means that the left circuit, with a resistance of 10 ohms, has a lower Q than the right circuit which has a resistance of 3 ohms.

EXAMPLE 12.4.5, DUBTUNE.CIR

The analysis of a circuit containing an RF transformer with a coefficient of coupling less than 1 is somewhat tedious, even at a single frequency.   To get a frequency response plot is not a task anyone would enjoy (or have time to do) by hand. Figure 12.31 illustrates such a circuit, which is fed by a current source.

One limitation of SPICE is apparent in this circuit; each node must have a DC path to ground so that a small-signal bias solution can be found.   This is true for any circuit, even if the bias solution is meaningless due to the fact that there are no DC sources in the circuit.   Resistor RDCPATH is added to the circuit so that each node will have a DC path to ground.   As shown in the input file, Fig. 12.32, RDCPATH has a value of 1E12 ohms, so it should not change the circuit performance very much.

Both primary and secondary windings of the transformer are resonant at 2 MHz, and the coefficient of coupling is greater than optimum coupling.   This means that the trans-

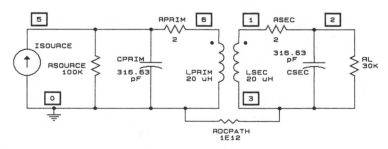

Figure 12.31   Double-Tuned RF Transformer.

```
DUBTUNE.CIR  DOUBLE-TUNED OVER-COUPLED RF TRANSFORMER
ISOURCE  0  5  AC  0.5M
RSOURCE 5  0  100K
CPRIM    5  0  316.63P
RPRIM    5  6  2
LPRIM    6  0  20U
KP-S     LPRIM LSEC  0.10
LSEC     1  3  20U
RSEC     1  2  2
CSEC     2  3  316.63P
RL       2  3  30K
RDCPATH 3  0  1T
.AC LIN  41  1.8MEG  2.2MEG
.PLOT  AC  VM(2,3) (0.3,5)
.OPTIONS NOPAGE
.END
```

Figure 12.32 Input File DUBTUNE.CIR.

former is overcoupled, and will have an undesirable frequency response.

Figure 12.33, the output file, shows the classic double-peaked response, where the output is a maximum at two frequencies, neither of which is the resonant frequency of 20 MHz.

```
********11/22/87******** Demo PSpice (May 1986) *******19:40:41********


DUBTUNE.CIR  DOUBLE-TUNED OVER-COUPLED RF TRANSFORMER


****    CIRCUIT DESCRIPTION


*****************************************************************************



ISOURCE       0  5 AC  0.5M
RSOURCE       5  0  100K
CPRIM    5  0  316.63P
RPRIM    5  6  2
LPRIM    6  0  20U
KP-S     LPRIM LSEC  0.10
LSEC     1  3  20U
RSEC     1  2  2
CSEC     2  3  316.63P
RL       2  3  30K
RDCPATH 3  0  1T
.AC LIN  41  1.8MEG  2.2MEG
.PLOT  AC  VM(2,3) (0.3,5)
.OPTIONS NOPAGE
.END
```

Figure 12.33 Output File DUBTUNE.OUT.

```
****      SMALL SIGNAL BIAS SOLUTION      TEMPERATURE =   27.000 DEG C
   NODE    VOLTAGE       NODE   VOLTAGE        NODE   VOLTAGE       NODE    VOLTAGE

 (  1)     .0000     (  2)     .0000      (  3)      .0000     (  5)      .0000
 (  6)     .0000

****      AC ANALYSIS                      TEMPERATURE =   27.000 DEG C
     FREQ       VM(2,3)

                   3.000D-01    6.062D-01    1.225D+00    2.475D+00    5.000D+00
             - - - - - - - - - - - - - - - - - - - - - - - - - - - - - - - - -
  1.800D+06  3.797D-01 .    *          :          :          :          :
  1.810D+06  4.324D-01 .      *        :          :          :          :
  1.820D+06  4.979D-01 .         *     :          :          :          :
  1.830D+06  5.813D-01 .            *. :          :          :          :
  1.840D+06  6.902D-01 .              . *         :          :          :
  1.850D+06  8.374D-01 .              :     *     :          :          :
  1.860D+06  1.045D+00 .              :         * :          :          :
  1.870D+06  1.356D+00 .              :          :*          :          :
  1.880D+06  1.856D+00 .              :          :     *     :          :
  1.890D+06  2.731D+00 .              :          :          : *         :
  1.900D+06  4.190D+00 .              :          :          :         * :
  1.910D+06  4.882D+00 .              :          :          :          :*
  1.920D+06  3.686D+00 .              :          :          :       *   :
  1.930D+06  2.691D+00 .              :          :          : *        :
  1.940D+06  2.118D+00 .              :          :       *   :          :
  1.950D+06  1.776D+00 .              :          :    *      :          :
  1.960D+06  1.559D+00 .              :          :  *        :          :
  1.970D+06  1.417D+00 .              :          : *         :          :
  1.980D+06  1.324D+00 .              :         .*           :          :
  1.990D+06  1.266D+00 .              :         .*           :          :
  2.000D+06  1.235D+00 .              :         *            :          :
  2.010D+06  1.229D+00 .              :         *            :          :
  2.020D+06  1.246D+00 .              :         *            :          :
  2.030D+06  1.288D+00 .              :         .*           :          :
  2.040D+06  1.360D+00 .              :          . *         :          :
  2.050D+06  1.472D+00 .              :          . *         :          :
  2.060D+06  1.641D+00 .              :          :    *      :          :
  2.070D+06  1.900D+00 .              :          :       *   :          :
  2.080D+06  2.310D+00 .              :          :          *.          :
  2.090D+06  2.986D+00 .              :          :          :  *        :
  2.100D+06  3.976D+00 .              :          :          :        *  :
  2.110D+06  4.314D+00 .              :          :          :         * .
  2.120D+06  3.190D+00 .              :          :          :      *    :
  2.130D+06  2.166D+00 .              :          :          * .        :
  2.140D+06  1.551D+00 .              :          :     *     :          :
  2.150D+06  1.174D+00 .              :          : *.        :          :
  2.160D+06  9.270D-01 .              :       *   :          :          :
  2.170D+06  7.555D-01 .              :   *       :          :          :
  2.180D+06  6.307D-01 .            .*            :          :          :
  2.190D+06  5.364D-01 .        *.               :          :          :
  2.200D+06  4.630D-01 .      *                   :          :          :
             - - - - - - - - - - - - - - - - - - - - - - - - - - - - - - - - -

        JOB CONCLUDED
        TOTAL JOB TIME          8.80
```

Figure 12.33 (Continued).

This is a good illustration of the concept of reflected impedance, in which the reactance of the secondary is reflected into the primary circuit and detunes the primary.

EXAMPLE 12.4.6, 2AC.CIR

A circuit with two AC sources is shown in Fig. 12.34. One is a voltage source, the other is a current source. The phasor voltage at node 5 is the unknown, over the frequency range of 1 KHz to 2 KHz.

The circuit requires a dummy resistor (as did Example 12.4.5, above) to provide a DC path to ground for node 9. In the small signal bias solution, the capacitor and the AC current source are both open circuits. RDUMMY has a value of 1 teraohm (1E12 ohm), which should have little effect on circuit behavior. The .AC control line in Fig. 12.35 will cause

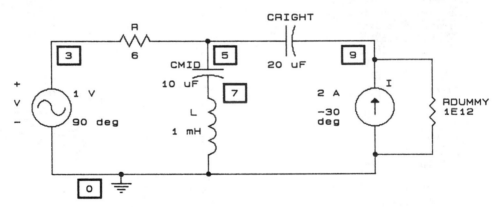

Figure 12.34 AC Circuit with Voltage and Current Source.

```
2AC.CIR           L,C,R CIRCUIT WITH V AND I SOURCE
V       3  0  AC 1 90
I       0  9  AC 2 -30
R       3  5  6
L       7  0  1E-3
CMID    5  7  10U
CRIGHT  5  9  20U
RDUMMY  9  0  1T
*       RDUMMY PROVIDES A DC PATH TO GROUND FOR NODE 9
.AC LIN 21 1K 2K
.PRINT AC VM(5) VP(5)
.PLOT AC VM(5) VP(5)
.OPTIONS NOPAGE
.END
```

Figure 12.35 Input File 2AC.CIR.

an analysis to occur at 21 frequencies linearly-spaced between 1 KHz and 2 KHz. Results of the AC analysis are both printed and plotted.

Figure 12.36 shows the output file 2AC.OUT. The graph indicates that the voltage at node 5 is a minimum around 1.6 KHz, and that the phase changes abruptly at the same frequency. SPICE always expresses phase angles such that the magnitude of the phase angle is 180 or less. Thus, what appears to be an abrupt change in phase angle may simply be due to how it is expressed by SPICE.

EXAMPLE 12.4.7, 2HP.CIR

Figure 12.37 shows two active filter circuits connected in parallel across an AC voltage source. Each circuit is a second-order high-pass filter; the first has Butterworth response,

```
********11/22/87******** Demo PSpice (May 1986) *******19:28:10********


2AC.CIR                L,C,R CIRCUIT WITH V AND I SOURCE


****    CIRCUIT DESCRIPTION


***********************************************************************


V      3  0  AC 1 90
I      0  9  AC 2 -30
R      3  5  6
L      7  0  1E-3
CMID   5  7  10U
CRIGHT 5  9  20U
RDUMMY 9  0  1T
*      RDUMMY PROVIDES A DC PATH TO GROUND FOR NODE 9
.AC LIN 21 1K 2K
.PRINT AC VM(5) VP(5)
.PLOT AC VM(5) VP(5)
.OPTIONS NOPAGE
.END
****    SMALL SIGNAL BIAS SOLUTION      TEMPERATURE =   27.000 DEG C

  NODE   VOLTAGE     NODE   VOLTAGE     NODE   VOLTAGE     NODE   VOLTAGE

 ( 3)    .0000     ( 5)    .0000     ( 7)    .0000     ( 9)    .0000
```

Figure 12.36 Output File 2AC.OUT.

```
****      AC ANALYSIS                        TEMPERATURE =   27.000 DEG C
    FREQ       VM(5)       VP(5)
  1.000E+03    9.789E+00   -5.761E+01
  1.050E+03    9.444E+00   -6.072E+01
  1.100E+03    9.032E+00   -6.414E+01
  1.150E+03    8.542E+00   -6.791E+01
  1.200E+03    7.960E+00   -7.205E+01
  1.250E+03    7.275E+00   -7.658E+01
  1.300E+03    6.479E+00   -8.151E+01
  1.350E+03    5.567E+00   -8.683E+01
  1.400E+03    4.543E+00   -9.249E+01
  1.450E+03    3.424E+00   -9.842E+01
  1.500E+03    2.236E+00   -1.045E+02
  1.550E+03    1.013E+00   -1.107E+02
  1.600E+03    2.035E-01    6.330E+01
  1.650E+03    1.377E+00    5.745E+01
  1.700E+03    2.477E+00    5.191E+01
  1.750E+03    3.483E+00    4.673E+01
  1.800E+03    4.387E+00    4.195E+01
  1.850E+03    5.186E+00    3.758E+01
  1.900E+03    5.887E+00    3.361E+01
  1.950E+03    6.496E+00    3.002E+01
  2.000E+03    7.025E+00    2.678E+01
****      AC ANALYSIS                        TEMPERATURE =   27.000 DEG C
LEGEND:
*: VM(5)
+: VP(5)
    FREQ      VM(5)
 (*)------------  1.000D-01    3.162D-01    1.000D+00    3.162D+00    1.000D+01
                  - - - - - - - - - - - - - - - - - - - - - - - - - - - - - - -
 (+)------------ -2.000D+02   -1.000D+02    .000D+00    1.000D+02    2.000D+02
                  - - - - - - - - - - - - - - - - - - - - - - - - - - - - - - -
 1.000D+03  9.789D+00 .              .      +        .             .             *
 1.050D+03  9.444D+00 .              .     +         .             .            *.
 1.100D+03  9.032D+00 .              .     +         .             .            *.
 1.150D+03  8.542D+00 .              .    +          .             .           *.
 1.200D+03  7.960D+00 .              .    +          .             .          *  .
 1.250D+03  7.275D+00 .              .   +           .             .         *   .
 1.300D+03  6.479D+00 .             . +             .             .        *    .
 1.350D+03  5.567D+00 .             . +             .             .      *      .
 1.400D+03  4.543D+00 .             .+              .             .    *        .
 1.450D+03  3.424D+00 .            +               .             . *           .
 1.500D+03  2.236D+00 .          +,               .         *    .             .
 1.550D+03  1.013D+00 .          +,               .    *         .             .
 1.600D+03  2.035D-01 .     *     .                .        +     .             .
 1.650D+03  1.377D+00 .           .                .     * +      .             .
 1.700D+03  2.477D+00 .           .                .     · + *    .             .
 1.750D+03  3.483D+00 .           .                .        +    .*             .
 1.800D+03  4.387D+00 .           .                .       +     .    *         .
 1.850D+03  5.186D+00 .           .                .      +      .      *       .
 1.900D+03  5.887D+00 .           .                .     +       .       *      .
 1.950D+03  6.496D+00 .           .                .     +       .        *     .
 2.000D+03  7.025D+00 .           .                .    +        .         *    .
                      - - - - - - - - - - - - - - - - - - - - - - - - - - - - - -
           JOB CONCLUDED
           TOTAL JOB TIME          6.40
```

Figure 12.36 (Continued).

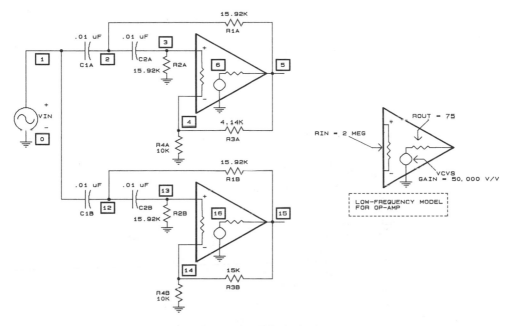

Figure 12.37 Two High-Pass Filters.

the second has Chebyshev response. The response is determined by the closed-loop gain of the operational amplifier.

A good AC model for an op-amp at low frequencies (where
the op-amp is not operated near its slew-rate and gain-
bandwidth product limits) is also shown in Fig. 12.37. The
model includes an input resistance which is rather large, and
a voltage-controlled voltage source which provides the open-
loop gain. The output voltage is the product of the differential input voltage and the open-loop gain.

The input file, Fig. 12.38, has comment lines to explain
the function of the various element and control lines. This
file would have been somewhat shorter if a subcircuit had
been used to model the op-amp and the breakpoint frequency-
determining components. The frequency of the source is logarithmically swept from 100 Hz to 10 KHz, with 20 frequencies
per decade.

The graph in the output file, Fig. 12.39, indicates that
the gain of the Chebyshev-response filter is higher than the
Butterworth-response filter at all frequencies, with a noticeable peak at 1.122 KHz. While not much quantitative data can
be obtained from this graph, it is of great use in understanding the difference between these two filters. A .PRINT control line could be added to the input file to obtain quantitative information.

```
2HP.CIR    TWO ACTIVE HIGH-PASS FILTERS IN PARALLEL
VIN      1  0  AC  1
*          CIRCUIT A FOLLOWS
C1A      1  2  .01U
C2A      2  3  .01U
R1A      5  2  15.92K
R2A      3  0  15.92K
R3A      5  4  4.14K
R4A      4  0  10K
*          OP-AMP MODEL FOR CIRCUIT A FOLLOWS
RINA     3  4  2MEG
EA       6  0  3  4  50K
ROUTA    6  5  75
*          CIRCUIT B FOLLOWS
C1B      1  12  .01U
C2B      12  13  .01U
R1B      15  12  15.92K
R2B      13  0  15.92K
R3B      15  14  15K
R4B      14  0  10K
*          OP-AMP MODEL FOR CIRCUIT B FOLLOWS
RINB     13  14  2MEG
EB       16  0  13  14  50K
ROUTB    16  15  75
*          CONTROL LINES FOLLOW
.AC  DEC  20  100  10K
.PLOT  AC  VDB(5)  VDB(15)  (-50,15)
.OPTIONS  NOPAGE
.END
```

Figure 12.38 Input File 2HP.CIR.

EXAMPLE 12.4.8, ACDIFAMP.CIR

A two-transistor BJT differential amplifier is shown in Fig. 12.40. It is fed by a voltage source at the base of Q1.

The input file, Fig. 12.41, includes two element lines for the transistors, but only one .MODEL line since both BJTs use the same model parameters. The model name is APPLE, which by itself means nothing until APPLE is defined in the .MODEL line. The model is of an NPN transistor, with a forward beta of 60 and significant junction capacitances representative of discrete devices, not integrated circuits.

A Bode plot is specified in the .AC and .PLOT control lines, in that the frequency is swept logarithmically from 100 Hz to 1 GHz and the output voltage (which is the gain, since the input voltage magnitude is one) is plotted in decibels.

```
********11/22/87******** Demo PSpice (May 1986) *******19:29:16********


2HP.CIR    TWO ACTIVE HIGH-PASS FILTERS IN PARALLEL


****      CIRCUIT DESCRIPTION


*************************************************************************



VIN      1  0  AC  1
*        CIRCUIT A FOLLOWS
C1A      1  2  .01U
C2A      2  3  .01U
R1A      5  2  15.92K
R2A      3  0  15.92K
R3A      5  4  4.14K
R4A      4  0  10K
*        OP-AMP MODEL FOR CIRCUIT A FOLLOWS
RINA     3  4  2MEG
EA       6  0  3  4  50K
ROUTA    6  5  75
*        CIRCUIT B FOLLOWS
C1B      1  12  .01U
C2B      12  13  .01U
R1B      15  12  15.92K
R2B      13  0  15.92K
R3B      15  14  15K
R4B      14  0  10K
*        OP-AMP MODEL FOR CIRCUIT B FOLLOWS
RINB     13  14  2MEG
EB       16  0  13  14  50K
ROUTB    16  15  75
*        CONTROL LINES FOLLOW
.AC  DEC  20  100  10K
.PLOT  AC  VDB(5)  VDB(15)  (-50,15)
.OPTIONS  NOPAGE
.END

****      SMALL SIGNAL BIAS SOLUTION     TEMPERATURE =   27.000 DEG C

  NODE   VOLTAGE     NODE   VOLTAGE     NODE   VOLTAGE     NODE   VOLTAGE


( 1)     .0000    ( 2)      .0000    ( 3)      .0000    ( 4)      .0000

( 5)     .0000    ( 6)      .0000    ( 12)     .0000    ( 13)     .0000

( 14)    .0000    ( 15)     .0000    ( 16)     .0000
```

Figure 12.39 Output File 2HP.OUT.

```
****    AC ANALYSIS                        TEMPERATURE =   27.000 DEG C

LEGEND:

*: VDB(5)
+: VDB(15)

     FREQ      VDB(5)
(*+)------------ -5.000D+01   -3.375D+01   -1.750D+01   -1.250D+00   1.500D+01
          - - - - - - - - - - - - - - - - - - - - - - - - - - -
 1.000D+02 -3.701D+01 .          *  .+            .            .            .
 1.122D+02 -3.502D+01 .           *. +            .            .            .
 1.259D+02 -3.302D+01 .           .*    +         .            .            .
 1.413D+02 -3.103D+01 .           . *    +        .            .            .
 1.585D+02 -2.904D+01 .           .  *    +       .            .            .
 1.778D+02 -2.706D+01 .           .   *    +      .            .            .
 1.995D+02 -2.508D+01 .           .    *    + .   .            .            .
 2.239D+02 -2.311D+01 .           .     *    +    .            .            .
 2.512D+02 -2.114D+01 .           .      *   . +  .            .            .
 2.818D+02 -1.919D+01 .           .       *. +    .            .            .
 3.162D+02 -1.725D+01 .           .       *    +  .            .            .
 3.548D+02 -1.532D+01 .           .        . *    +           .            .
 3.981D+02 -1.343D+01 .           .         . *    +          .            .
 4.467D+02 -1.157D+01 .           .          . *    + .       .            .
 5.012D+02 -9.752D+00 .           .           .    *    +,    .            .
 5.623D+02 -8.001D+00 .           .           .     *    .+   .            .
 6.310D+02 -6.335D+00 .           .           .      *      +  .           .
 7.079D+02 -4.776D+00 .           .           .       *      .  +          .
 7.943D+02 -3.352D+00 .           .           .        *.       +          .
 8.913D+02 -2.086D+00 .           .           .         *.        .  +     .
 1.000D+03 -9.951D-01 .           .           .          *        .      +.
 1.122D+03 -8.651D-02 .           .           .           .*      .      +.
 1.259D+03  6.465D-01 .           .           .           . *     .      +.
 1.413D+03  1.221D+00 .           .           .           . *     .       +.
 1.585D+03  1.662D+00 .           .           .           . *     .    +  .
 1.778D+03  1.995D+00 .           .           .           .  *     +     .
 1.995D+03  2.244D+00 .           .           .           . *      +     .
 2.239D+03  2.430D+00 .           .           .           . *      +     .
 2.512D+03  2.568D+00 .           .           .           . *     +      .
 2.818D+03  2.672D+00 .           .           .           . *     +      .
 3.162D+03  2.749D+00 .           .           .           . *     +      .
 3.548D+03  2.808D+00 .           .           .           . *     +      .
 3.981D+03  2.853D+00 .           .           .           . *     +      .
 4.467D+03  2.887D+00 .           .           .           . *     +      .
 5.012D+03  2.914D+00 .           .           .           . *     +      .
 5.623D+03  2.934D+00 .           .           .           . *     +      .
 6.310D+03  2.950D+00 .           .           .           . *     +      .
 7.079D+03  2.963D+00 .           .           .           . *    +       .
 7.943D+03  2.972D+00 .           .           .           . *    +       .
 8.913D+03  2.980D+00 .           .           .           . *    +       .
 1.000D+04  2.986D+00 .           .           .           . *    +       .
          - - - - - - - - - - - - - - - - - - - - - - - - - - -

        JOB CONCLUDED

        TOTAL JOB TIME           12.30
```

Figure 12.39 (Continued).

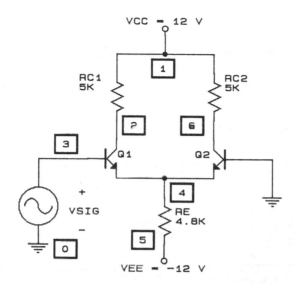

Figure 12.40   BJT Differential Amplifier.

It is important to note that the AC analysis uses small-signal models for all devices.   The input voltage of 1 V is hardly a small-signal input, and it causes a great deal more than a small-signal output.   The output voltage in Fig. 12.42 shows an output voltage in the passband of 40.90 dBV, or 111 V.   This is a ridiculous result and could not happen in the actual circuit because the power supplies would limit the output to about 12 Vpp.   However, it is convenient to use 1 V as the input magnitude, since the output voltage is then the same as the circuit gain.

```
ACDIFAMP.CIR    BJT DIFFERENTIAL AMPLIFIER, AC ANALYSIS
VSIG    3   0   AC  1
VCC     1   0   DC  12
VEE     5   0   -12
Q1      2   3   4   APPLE
Q2      6   0   4   APPLE
.MODEL APPLE NPN(BF=60  CJC=16P  CJE=30P)
RC1     1   2   5K
RC2     1   6   5K
RE      4   5   4.8K
.AC  DEC  5  100  1G
.PLOT  AC  VDB(6)  (-10,50)  VP(6)  (-90,0)
.OPTIONS NOPAGE
.OP
.END
```

Figure 12.41   Input File ACDIFAMP.CIR.

```
********11/22/87******** Demo PSpice (May 1986) ********19:30:17********

ACDIFAMP.CIR    BJT DIFFERENTIAL AMPLIFIER, AC ANALYSIS

****    CIRCUIT DESCRIPTION

*************************************************************************


VSIG   3   0   AC   1
VCC    1   0   DC   12
VEE    5   0   -12
Q1     2   3   4   APPLE
Q2     6   0   4   APPLE
.MODEL APPLE NPN(BF=60   CJC=16P   CJE=30P)
RC1    1   2   5K
RC2    1   6   5K
RE     4   5   4.8K
.AC  DEC  5  100  1G
.PLOT  AC  VDB(6)  (-10,50)  VP(6)  (-90,0)
.OPTIONS NOPAGE
.OP
.END

****    BJT MODEL PARAMETERS


           APPLE

TYPE       NPN

IS      1.00D-16

BF        60.000

NF         1.000

BR         1.000

NR         1.000

CJE     3.00D-11

CJC     1.60D-11
```

Figure 12.42 Output File ACDIFAMP.OUT.

```
****     SMALL SIGNAL BIAS SOLUTION      TEMPERATURE =   27.000 DEG C

  NODE    VOLTAGE     NODE    VOLTAGE    NODE    VOLTAGE    NODE    VOLTAGE


 (  1)    12.0000    (  2)    6.2509    (  3)     .0000    (  4)    -.7778

 (  5)   -12.0000    (  6)    6.2509
```

```
    VOLTAGE SOURCE CURRENTS

    NAME        CURRENT


    VSIG      -1.916D-05

    VCC       -2.300D-03

    VEE        2.338D-03


    TOTAL POWER DISSIPATION    5.57D-02  WATTS
```

```
****      OPERATING POINT INFORMATION      TEMPERATURE =   27.000 DEG C



**** BIPOLAR JUNCTION TRANSISTORS



               Q1          Q2

MODEL        APPLE       APPLE
IB           1.92E-05    1.92E-05
IC           1.15E-03    1.15E-03
VBE          .778        .778
VBC          -6.25       -6.25
VCE          7.03        7.03
BETADC       60.0        60.0
GM           4.45E-02    4.45E-02
RPI          1.35E+03    1.35E+03
RX           .00E+00     .00E+00
RO           1.00E+12    1.00E+12
CPI          5.11E-11    5.11E-11
CMU          7.66E-12    7.66E-12
CBX          .00E+00     .00E+00
CCS          .00E+00     .00E+00
BETAAC       60.0        60.0
FT           1.20E+08    1.20E+08
```

Figure 12.42 (Continued).

```
****    AC ANALYSIS                      TEMPERATURE =   27.000 DEG C

LEGEND:

*: VDB(6)
+: VP(6)

     FREQ      VDB(6)

(*)------------- -1.000D+01    5.000D+00    2.000D+01    3.500D+01    5.000D+01
                             - - - - - - - - - - - - - - - - - - - - - - - - -

(+)------------- -9.000D+01   -6.750D+01   -4.500D+01   -2.250D+01    .000D+00
                             - - - - - - - - - - - - - - - - - - - - - - - - -

1.000D+02   4.090D+01 .           .            .            .        *       +
1.585D+02   4.090D+01 .           .            .            .        *       +
2.512D+02   4.090D+01 .           .            .            .      `*       +
3.981D+02   4.090D+01 .           .            .            .        *       +
6.310D+02   4.090D+01 .           .            .            .        *      .+
1.000D+03   4.090D+01 .           .            .            .        *       +
1.585D+03   4.090D+01 .           .            .            .        *       +
2.512D+03   4.090D+01 .           .            .            .        *       +
3.981D+03   4.090D+01 .           .            .            .        *       +
6.310D+03   4.090D+01 .           .            .            .        *       +
1.000D+04   4.090D+01 .           .            .            "        *       +
1.585D+04   4.090D+01 .           .            .            .        *       +
2.512D+04   4.090D+01 .           .            .            .        *       +
3.981D+04   4.090D+01 .           .            .            .        *       +
6.310D+04   4.090D+01 .           .            .            .        *      +.
1.000D+05   4.089D+01 .           .            .            .        *      +.
1.585D+05   4.089D+01 .           .            .            "        *      +.
2.512D+05   4.088D+01 .           .            .            "        *    +  .
3.981D+05   4.086D+01 .           .            .            "        *   +   .
6.310D+05   4.080D+01 .           .            .            .       * +     .
1.000D+06   4.065D+01 .           .            .            "       X       .
1.585D+06   4.031D+01 .           .            .            .+      *       "
2.512D+06   3.955D+01 .           .            .         +  .    *          "
3.981D+06   3.807D+01 .           .            .+          .   *            "
6.310D+06   3.571D+01 .           .         +  .          .*              "
1.000D+07   3.258D+01 .           .     +      .         * .               "
1.585D+07   2.899D+01 .       +    .          .        *                  "
2.512D+07   2.516D+01 .    +       .          .     *                      "
3.981D+07   2.123D+01 .  +         .         .*                            "
6.310D+07   1.726D+01 . +          .     *  .                              "
1.000D+08   1.327D+01 .+          .     *    .                             "
1.585D+08   9.283D+00 .+          .   *      .                             "
2.512D+08   5.289D+00 .+        *  .          .                             "
3.981D+08   1.292D+00 +        *   .          .                             "
6.310D+08  -2.707D+00 +     *      .          .          "                 "
1.000D+09  -6.706D+00 +  *         .          .          "                 "
                             - - - - - - - - - - - - - - - - - - - - - - - - -

        JOB CONCLUDED

        TOTAL JOB TIME        10.40
```

Figure 12.42 (Continued).

If the input were specified as 1 KV, the output would simply be 111 KV. Clearly, you must be careful when interpreting results of an AC analysis to be sure that the numbers are reasonable. If you wanted to see the effects of non-linearities, such as clipping of the output due to the transistors becoming cutoff and saturated, a transient analysis would have to be performed. The transient analysis is not limited to small-signal models as the AC analysis is.

The Bode plot shows the half-power frequency to be just greater than 3.981 MHz, as indicated by the 3 dB drop in VDB(6) and the phase angle of -45 degrees near that frequency.

12.5 TRANSIENT ANALYSIS

EXAMPLE 12.5.1, TRIANG.CIR

A capacitor and resistor in series are connected to a triangle voltage source, as illustrated in Fig. 12.43. The triangle is made with the piece-wise linear (PWL) function, and varies between 0 V and 1 V, with a period of 2 ms.

Figure 12.44 lists input file TRIANG.CIR, which defines the triangle voltage from 0 ms to 4 ms only. Note that there

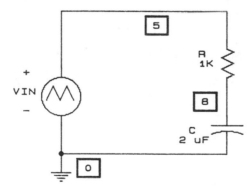

Figure 12.43 R-C Circuit, Triangle Input.

```
TRIANG.CIR    PIECEWISE LINEAR VOLTAGE INTO R-C LOWPASS FILTER
VIN     5   0   PWL(0 0  1M 1  2M 0  3M 1  4M 0)
R       5   8   1K
C       8   0   2U
.TRAN  .1M  4M
.PLOT  TRAN  V(5)  (0,1)  V(8)  (0,.5)
.OPTIONS NOPAGE
.END
```

Figure 12.44 Input File TRIANG.CIR.

are five time-voltage pairs in the PWL definition. The transient analysis is from 0 ms to 4 ms, using time steps of 0.1 ms. The .PLOT control line specifies that the input voltage, V(5), and the capacitor voltage, V(8), will be plotted with the plot limits shown.

An examination of the graph of the capacitor voltage in the output file, Fig. 12.45, indicates that the capacitor voltage did not reach steady-state in 4 ms. It was last seen heading to an average value of 0.5 V, which is the average value of the triangle input voltage.

For analyses that require many cycles of a periodic triangle waveform, using a PWL source can become rather tedious. A better way to generate a periodic triangle, pulse, sawtooth, etc. waveform is to use the PULSE function. When doing so, it is important to use a tiny (compared to the period) but non-zero value of pulse width. The rise time and fall time are set to equal half the period of the triangle waveform. A periodic triangle waveform identical to the PWL triangle in this example could be produced by the following control line

    VIN 5 0 PULSE(0 1 0 1M 1M 1N 2M)

Reading from left to right in the parentheses above, the initial pulse value is 0 V, the pulsed value is 1 V, the delay

```
********11/22/87******** Demo PSpice (May 1986) *******19:56:13********

TRIANG.CIR   PIECEWISE LINEAR VOLTAGE INTO R-C LOWPASS FILTER

****    CIRCUIT DESCRIPTION

*************************************************************************
VIN   5  0  PWL(0 0  1M 1  2M 0  3M 1  4M 0)
R     5  8  1K
C     8  0  2U
.TRAN .1M  4M
.PLOT TRAN V(5)  (0,1) V(8) (0,.5)
.OPTIONS NOPAGE
.END

****    INITIAL TRANSIENT SOLUTION      TEMPERATURE =   27.000 DEG C

 NODE   VOLTAGE    NODE   VOLTAGE

( 5)    .0000   ( 8)     .0000
```

Figure 12.45 Output File TRIANG.OUT

```
****      TRANSIENT ANALYSIS                 TEMPERATURE =   27.000 DEG C
LEGEND:
*: V(5)
+: V(8)
    TIME       V(5)
(*)------------    .000D+00     2.500D-01    5.000D-01    7.500D-01    1.000D+00
                   - - - - - - - - - - - - - - - - - - - - - - - - - - - - - -
(+)------------    .000D+00     1.250D-01    2.500D-01    3.750D-01    5.000D-01
                   - - - - - - - - - - - - - - - - - - - - - - - - - - - - - -
  .000D+00    .000D+00 X          .            .            .            .
 1.000D-04   1.000D-01 +    *     .            .            .            .
 2.000D-04   2.000D-01 .+      *  .            .            .            .
 3.000D-04   3.000D-01 .  +       . *          .            .            .
 4.000D-04   4.000D-01 .    +     .        *   .            .            .
 5.000D-04   5.000D-01 .      +   .            *            .            .
 6.000D-04   6.000D-01 .        + .            .     *      .            .
 7.000D-04   7.000D-01 .         +.            .            *  .         .
 8.000D-04   8.000D-01 .          . +          .            .  *         .
 9.000D-04   9.000D-01 .          .    +       .            .        *   .
 1.000D-03   1.000D+00 .          .        +   .            .            *
 1.100D-03   9.000D-01 .          .            +            .        *   .
 1.200D-03   8.000D-01 .          .            .   +        .     *      .
 1.300D-03   7.000D-01 .          .            .     +      *            .
 1.400D-03   6.000D-01 .          .            .        *  +.            .
 1.500D-03   5.000D-01 .          .            .   *         +           .
 1.600D-03   4.000D-01 .          .        *   .            +            .
 1.700D-03   3.000D-01 .          .    *       .            +            .
 1.800D-03   2.000D-01 .      *    .            .           +            .
 1.900D-03   1.000D-01 .    *      .            .           +            .
 2.000D-03    .000D+00 *          .            .         +  .            .
 2.100D-03   1.000D-01 .  *       .            .        +   .            .
 2.200D-03   2.000D-01 .     *    .            .      +     .            .
 2.300D-03   3.000D-01 .          . *          .     +      .            .
 2.400D-03   4.000D-01 .          .     *      .    +       .            .
 2.500D-03   5.000D-01 .          .            * +         .            .
 2.600D-03   6.000D-01 .          .            .  *+        .            .
 2.700D-03   7.000D-01 .          .            .       + *  .            .
 2.800D-03   8.000D-01 .          .            .       +   .*            .
 2.900D-03   9.000D-01 .          .            .           +  .    *     .
 3.000D-03   1.000D+00 .          .            .           .   +        *
 3.100D-03   9.000D-01 .          .            .           .      +  *   .
 3.200D-03   8.000D-01 .          .            .           .   *   +     .
 3.300D-03   7.000D-01 .          .            .        *  .        +    .
 3.400D-03   6.000D-01 .          .            .      *    .         +   .
 3.500D-03   5.000D-01 .          .            *          .         +   .
 3.600D-03   4.000D-01 .          .     *      .          .         +   .
 3.700D-03   3.000D-01 .          . *          .          .         +   .
 3.800D-03   2.000D-01 .      *    .            .          .         +   .
 3.900D-03   1.000D-01 .    *      .            .          .      +      .
 4.000D-03    .000D+00 *          .            .          .   +         .
                   - - - - - - - - - - - - - - - - - - - - - - - - - - - - - -
          JOB CONCLUDED
          TOTAL JOB TIME          9.60
```

Figure 12.45 (Continued).

time is 0, the rise time is 1 ms, the fall time is 1 ms, the pulse width is 1 ns (one millionth the rise or fall time) and the period is 2 ms.   After 1000 cycles or so, the 1 ns pulse width, which should be 0, will start to introduce a slight error (a cumulative lengthening of the period by 0.0001% each cycle).   However, the error is so slight that it doesn't much matter for any practical purpose.

EXAMPLE 12.5.2, LC.CIR

An inductor and capacitor in parallel (tank circuit) are connected to a pulse current source, as illustrated in Fig. 12.46.   The tank circuit has a resonant frequency of 1007 Hz. This is a circuit that can be realized on paper only, as there are no losses in the circuit.   The pulse is very short, with a magnitude of 1 A.

The input file in Fig. 12.47 shows the PWL current source pulse to have a rise time, pulse width and fall time of 0.1 ms each.   The pulse is delayed from time zero by 1 ms.

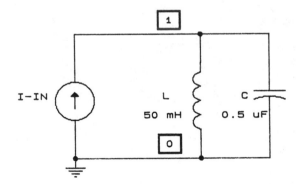

Figure 12.46 Current Pulse into Tank Circuit.

```
LC.CIR    PARALLEL L-C, NO LOSS, PULSE INPUT
I-IN    0  1  PWL(0  0  1M  0  1.1M  1  1.2M  1  1.3M  0)
L       1  0  50M
C       1  0  0.5U
.TRAN  .1M  4M
.PLOT  TRAN  V(1)  (-400,400)  I(I-IN)  (0,2)
.OPTIONS  NOPAGE
.END
```

Figure 12.47 Input File LC.CIR.

Examination of the output file, Fig. 12.48, shows the current pulse plotted with plus signs and the voltage across the tank circuit plotted with asterisks.   The pulse starts at 1 ms, and the tank voltage begins ringing at the same time. After the pulse ends at 1.3 ms, the tank voltage continues to ring with constant amplitude.   This is due to the total absence of loss in the circuit; the tank will ring for as long as you care to do the transient analysis.

EXAMPLE 12.5.3, EKG-LP.CIR

A piece-wise linear source can be used to make any arbitrary waveform whatsoever, providing you have the patience to put all the information on the element line.   This example contains an electrocardiogram (EKG) voltage over a time interval of 500 ms, with data points every 5 ms.   The PWL independent source element "line" actually requires 16 lines, using the plus symbol to indicate a continuation of the previous line. The circuit, Fig. 12.49, is a single-pole R-C low-pass filter

```
********11/20/87******** Demo PSpice (May 1986) *******15:56:18**xxxxxx*


LC.CIR   PARALLEL L-C, NO LOSS, PULSE INPUT


****    CIRCUIT DESCRIPTION


**********************************************************************



I-IN   0  1  PWL(0  0  1M  0  1.1M  1  1.2M  1  1.3M  0)
L      1  0  50M
C      1  0  0.5U
.TRAN  .1M  4M
.PLOT  TRAN  V(1)  (-400,400)  I(I-IN)  (0,2)
.OPTIONS  NOPAGE
.END

****    INITIAL TRANSIENT SOLUTION     TEMPERATURE =   27.000 DEG C

 NODE   VOLTAGE


(  1)    .0000
```

Figure 12.48 Output File LC.OUT.

```
****     TRANSIENT ANALYSIS              TEMPERATURE =   27.000 DEG C
LEGEND:
*: V(1)
+: I(I-IN)
     TIME        V(1)
(*)------------- -4.000D+02    -2.000D+02    .000D+00     2.000D+02     4.000D+02
                 - - - - - - - - - - - - - - - - - - - - - - - - - - - -
(+)-------------  .000D+00     5.000D-01    1.000D+00    1.500D+00     2.000D+00
                 - - - - - - - - - - - - - - - - - - - - - - - - - - - -
   .000D+00    .000D+00  +              .            *            .            .
  1.000D-04    .000D+00  +              .            *            .            .
  2.000D-04    .000D+00  +              .            *            .            .
  3.000D-04    .000D+00  +              .            *            .            .
  4.000D-04    .000D+00  +              .            *            .            .
  5.000D-04    .000D+00  +              .            *            .            .
  6.000D-04    .000D+00  +              .            *            .            .
  7.000D-04    .000D+00  +              .            *            .            .
  8.000D-04    .000D+00  +              .            *            .            .
  9.000D-04    .000D+00  +              .            *            .            .
  1.000D-03    .000D+00  +              .            *            .            .
  1.100D-03   9.643D+01  .              .            +        *   .            .
  1.200D-03   2.518D+02  .              .            +            .  *         .
  1.300D-03   2.136D+02  +              .            .            .*           .
  1.400D-03  -6.996D-01  +              .            *            .            .
  1.500D-03  -2.064D+02  +         *    .            .            .            .
  1.600D-03  -3.349D+02  +    *         .            .            .            .
  1.700D-03  -3.431D+02  +    *         .            .            .            .
  1.800D-03  -2.220D+02  +         *.   .            .            .            .
  1.900D-03  -1.513D+01  +              .         *. .            .            .
  2.000D-03   1.936D+02  +              .            .            *            .
  2.100D-03   3.339D+02  +              .            .            .        *   .
  2.200D-03   3.505D+02  +              .            .            .         *  .
  2.300D-03   2.302D+02  +              .            .            .   *        .
  2.400D-03   2.988D+01  +              .            .  *         .            .
  2.500D-03  -1.830D+02  +           .* .            .            .            .
  2.600D-03  -3.301D+02  +      *       .            .            .            .
  2.700D-03  -3.476D+02  +    *         .            .            .            .
  2.800D-03  -2.407D+02  +         *    .            .            .            .
  2.900D-03  -4.567D+01  +              .       *    .            .            .
  3.000D-03   1.695D+02  +              .            .        *   .            .
  3.100D-03   3.173D+02  +              .            .            .      *     .
  3.200D-03   3.493D+02  +              .            .            .        *   .
  3.300D-03   2.549D+02  +              .            .            .    *       .
  3.400D-03   6.299D+01  +              .            .     *      .            .
  3.500D-03  -1.523D+02  +           .   *           .            .            .
  3.600D-03  -3.095D+02  +      *       .            .            .            .
  3.700D-03  -3.559D+02  +    *         .            .            .            .
  3.800D-03  -2.688D+02  +        *     .            .            .            .
  3.900D-03  -7.736D+01  +              .      *     .            .            .
  4.000D-03   1.437D+02  +              .            .       *    .            .
                 - - - - - - - - - - - - - - - - - - - - - - - - - - - -

                JOB CONCLUDED
                TOTAL JOB TIME          10.60
```

Figure 12.48 (Continued).

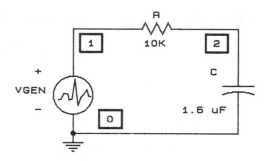

Figure 12.49 EKG into Low-Pass Filter.

connected to the EKG voltage. It has a breakpoint frequency
of 10 Hz.

The input file in Fig. 12.50 specifies that a transient
analysis occur, with a time step of 10 ms, and that the EKG
voltage and the filtered EKG be plotted versus time. The optional plot limits were chosen to give maximum size to the
EKG plot, and to make the filtered EKG plot have the same
scale.

Figure 12.51 shows the output file EKG-LP.OUT. The effect of low-pass filtering on a complex waveform is very apparent, in that the filtered waveform is smoothed consider-

```
EKG-LP.CIR  ELECTRO-CARDIOGRAM VOLTAGE INTO LOW-PASS FILTER
VGEN 1 Ø PWL(Ø -1 5M 2 10M 2 15M 2 20M 4 25M 5 30M 4
+ 35M 8 40M 8 45M 7 50M 8 55M 8 60M 6 65M 4 70M 3 75M -1
+ 80M -4 85M -4 90M -6 95M -9 100M -9 105M -7 110M -8
+ 115M -9 120M -7 125M -7 130M -8 135M -6 140M -7
+ 145M -9 150M -9 155M -8 160M -10 165M -9 170M -2
+ 175M 9 180M 25 185M 44 190M 54 195M 49 200M 34 205M 12
+ 210M -10 215M -25 220M -30 225M -30 230M -26 235M -20
+ 240M -14 245M -10 250M -5 255M -2 260M -3 265M -3
+ 270M -1 275M -1 280M -2 285M Ø 290M 1 295M Ø 300M 1
+ 305M 3 310M 2 315M 2 320M 4 325M 5 330M 3 335M 6
+ 340M 7 345M 7 350M 9 355M 11 360M 11 365M 12 370M 15
+ 375M 18 380M 18 385M 23 390M 25 395M 25 400M 28
+ 405M 32 410M 33 415M 35 420M 38 425M 38 430M 37
+ 435M 39 440M 36 445M 34 450M 31 455M 26 460M 22
+ 465M 19 470M 14 475M 10 480M 7 485M 5 490M 2
+ 495M 1 500M 1)
R       1 2 10K
C       2 Ø 1.6U
.TRAN 10M 500M
.PLOT TRAN V(1,Ø)  (60,-30)
.PLOT TRAN V(2,Ø) (60,-30)
.OPTIONS NOPAGE
.END
```

Figure 12.50 Input File EKG-LP.CIR.

ably and has a much smaller peak-to-peak value between 180 ms and 220 ms.

EXAMPLE 12.5.4, PWRSUPLI.CIR

An interesting circuit to study with transient analysis is a half-wave rectifier with a capacitor filter. The inrush current at the time of turn-on is not easily measured in the lab-

```
********11/24/87******** Demo PSpice (May 1986) ******* 9:05:08********


EKG-LP.CIR  ELECTRO-CARDIOGRAM VOLTAGE INTO LOW-PASS FILTER


****     CIRCUIT DESCRIPTION


***************************************************************************

VGEN 1 0 PWL(0 -1 5M 2 10M 2 15M 2 20M 4 25M 5 30M 4
+ 35M 8 40M 8 45M 7 50M 8 55M 8 60M 6 65M 4 70M 3 75M -1
+ 80M -4 85M -4 90M -6 95M -9 100M -9 105M -7 110M -8
+ 115M -9 120M -7 125M -7 130M -8 135M -6 140M -7
+ 145M -9 150M -9 155M -8 160M -10 165M -9 170M -2
+ 175M 9 180M 25 185M 44 190M 54 195M 49 200M 34 205M 12
+ 210M -10 215M -25 220M -30 225M -30 230M -26 235M -20
+ 240M -14 245M -10 250M -5 255M -2 260M -3 265M -3
+ 270M -1 275M -1 280M -2 285M 0 290M 1 295M 0 300M 1
+ 305M 3 310M 2 315M 2 320M 4 325M 5 330M 3 335M 6
+ 340M 7 345M 7 350M 9 355M 11 360M 11 365M 12 370M 15
+ 375M 18 380M 18 385M 23 390M 25 395M 25 400M 28
+ 405M 32 410M 33 415M 35 420M 38 425M 38 430M 37
+ 435M 39 440M 36 445M 34 450M 31 455M 26 460M 22
+ 465M 19 470M 14 475M 10 480M 7 485M 5 490M 2
+ 495M 1 500M 1)
R     1  2  10K
C     2  0  1.6U
.TRAN 10M 500M
.PLOT TRAN V(1,0)  (60,-30)
.PLOT TRAN V(2,0)  (60,-30)
.OPTIONS NOPAGE
.END

****     INITIAL TRANSIENT SOLUTION     TEMPERATURE =   27.000 DEG C

  NODE   VOLTAGE     NODE   VOLTAGE


 ( 1)   -1.0000    ( 2)   -1.0000
```

Figure 12.51 Output File EKG-LP.OUT.

```
****    TRANSIENT ANALYSIS                  TEMPERATURE =   27.000 DEG C
     TIME      V(1)

                         -3.000D+01   -7.500D+00    1.500D+01    3.750D+01    6.000D+01
 ------------------------------------------------------------------------------------
  .000D+00 -1.000D+00 .                    .        *        .            .            .
 1.000D-02  2.000D+00 .                    .         *       .            .            .
 2.000D-02  4.000D+00 .                    .          *      .            .            .
 3.000D-02  4.000D+00 .                    .          *      .            .            .
 4.000D-02  8.000D+00 .                    .            *    .            .            .
 5.000D-02  8.000D+00 .                    .            *    .            .            .
 6.000D-02  6.000D+00 .                    .           *     .            .            .
 7.000D-02  3.000D+00 .                    .         *       .            .            .
 8.000D-02 -4.000D+00 .                    . *             .            .            .
 9.000D-02 -6.000D+00 .                    .*             .            .            .
 1.000D-01 -9.000D+00 .                  *,             .            .            .
 1.100D-01 -8.000D+00 .                   *             .            .            .
 1.200D-01 -7.000D+00 .                   *             .            .            .
 1.300D-01 -8.000D+00 .                   *             .            .            .
 1.400D-01 -7.000D+00 .                   *             .            .            .
 1.500D-01 -9.000D+00 .                  *.             .            .            .
 1.600D-01 -1.000D+01 .                  *.             .            .            .
 1.700D-01 -2.000D+00 .                    .  *          .            .            .
 1.800D-01  2.500D+01 .                    .             .    *        .            .
 1.900D-01  5.400D+01 .                    .             .            .          * .
 2.000D-01  3.400D+01 .                    .             .            * .            .
 2.100D-01 -1.000D+01 .                  *.             .            .            .
 2.200D-01 -3.000D+01 *                    .             .            .            .
 2.300D-01 -2.600D+01 . *                  .             .            .            .
 2.400D-01 -1.400D+01 .          *          .            .            .            .
 2.500D-01 -5.000D+00 .                 .*             .            .            .
 2.600D-01 -3.000D+00 .                    . *           .            .            .
 2.700D-01 -1.000D+00 .                    .  *          .            .            .
 2.800D-01 -2.000D+00 .                    . *           .            .            .
 2.900D-01  1.000D+00 .                    .    *        .            .            .
 3.000D-01  1.000D+00 .                    .    *        .            .            .
 3.100D-01  2.000D+00 .                    .    *        .            .            .
 3.200D-01  4.000D+00 .                    .      *      .            .            .
 3.300D-01  3.000D+00 .                    .     *       .            .            .
 3.400D-01  7.000D+00 .                    .       *     .            .            .
 3.500D-01  9.000D+00 .                    .        *    .            .            .
 3.600D-01  1.100D+01 .                    .         *.  .            .            .
 3.700D-01  1.500D+01 .                    .          *  .            .            .
 3.800D-01  1.800D+01 .                    .           * .            .            .
 3.900D-01  2.500D+01 .                    .             .    *        .            .
 4.000D-01  2.800D+01 .                    .             .     *       .            .
 4.100D-01  3.300D+01 .                    .             .       *     .            .
 4.200D-01  3.800D+01 .                    .             .            * .            .
 4.300D-01  3.700D+01 .                    .             .            * .            .
 4.400D-01  3.600D+01 .                    .             .            *.            .
 4.500D-01  3.100D+01 .                    .             .          *  .            .
 4.600D-01  2.200D+01 .                    .             .    *        .            .
 4.700D-01  1.400D+01 .                    .          *. .            .            .
 4.800D-01  7.000D+00 .                    .       *     .            .            .
 4.900D-01  2.000D+00 .                    .    *        .            .            .
 5.000D-01  1.000D+00 .                    .    *        .            .            .
 ------------------------------------------------------------------------------------
```

Figure 12.51 (Continued).

```
****    TRANSIENT ANALYSIS              TEMPERATURE =   27.000 DEG C
     TIME       V(2)
                    -3.000D+01   -7.500D+00    1.500D+01    3.750D+01    6.000D+01
                   - - - - - - - - - - - - - - - - - - - - - - - - - - - - -
 .000D+00 -1.000D+00 .             .           *          .            .          .
1.000D-02  1.143D-01 .             .           *          .            .          .
2.000D-02  1.272D+00 .             .            *         .            .          .
3.000D-02  2.771D+00 .             .             *        .            .          .
4.000D-02  4.830D+00 .             .              *       .            .          .
5.000D-02  6.072D+00 .             .               *      .            .          .
6.000D-02  6.687D+00 .             .               *      .            .          .
7.000D-02  5.483D+00 .             .               *      .            .          .
8.000D-02  2.420D+00 .             .             *        .            .          .
9.000D-02 -8.459D-01 .             .          *           .            .          .
1.000D-01 -4.357D+00 .             . *                    .            .          .
1.100D-01 -5.912D+00 .             .*                     .            .          .
1.200D-01 -6.971D+00 .            *.                       .            .          .
1.300D-01 -7.125D+00 .            *.                       .            .          .
1.400D-01 -6.929D+00 .            *.                       .            .          .
1.500D-01 -7.705D+00 .            *.                       .            .          .
1.600D-01 -8.217D+00 .            *.                       .            .          .
1.700D-01 -7.686D+00 .            *.                       .            .          .
1.800D-01  1.307D+00 .             .         *            .            .          .
1.900D-01  2.079D+01 .             .                    *  .            .          .
2.000D-01  3.224D+01 .             .                       .         *  .          .
2.100D-01  2.178D+01 .             .                     *  .            .          .
2.200D-01  7.335D-01 .             .          *           .            .          .
2.300D-01 -1.298D+01 .        *    .                       .            .          .
2.400D-01 -1.596D+01 .      *      .                       .            .          .
2.500D-01 -1.286D+01 .       *     .                       .            .          .
2.600D-01 -8.230D+00 .           * .                       .            .          .
2.700D-01 -5.518D+00 .            .*                       .            .          .
2.800D-01 -3.558D+00 .            . *                      .            .          .
2.900D-01 -1.950D+00 .             .*                      .            .          .
3.000D-01 -8.096D-01 .             . *                     .            .          .
3.100D-01  6.344D-01 .             .    *                  .            .          .
3.200D-01  1.551D+00 .             .    *                  .            .          .
3.300D-01  2.780D+00 .             .     *                 .            .          .
3.400D-01  4.137D+00 .             .      *                .            .          .
3.500D-01  5.751D+00 .             .       *               .            .          .
3.600D-01  8.004D+00 .             .        *              .            .          .
3.700D-01  1.019D+01 .             .          * .          .            .          .
3.800D-01  1.354D+01 .             .           *.          .            .          .
3.900D-01  1.775D+01 .             .            .*         .            .          .
4.000D-01  2.154D+01 .             .            .    *     .            .          .
4.100D-01  2.617D+01 .             .            .       *  .            .          .
4.200D-01  3.051D+01 .             .            .         *.            .          .
4.300D-01  3.385D+01 .             .            .          . *          .          .
4.400D-01  3.563D+01 .             .            .          .  *.         .          .
4.500D-01  3.464D+01 .             .            .          . *           .          .
4.600D-01  3.053D+01 .             .            .          *.            .          .
4.700D-01  2.474D+01 .             .            .      *   .            .          .
4.800D-01  1.784D+01 .             .            .*         .            .          .
4.900D-01  1.163D+01 .             .         *.            .            .          .
5.000D-01  6.785D+00 .             .       *  .            .            .          .
                   - - - - - - - - - - - - - - - - - - - - - - - - - - - - -
         JOB CONCLUDED
         TOTAL JOB TIME            47.20
```

Figure 12.51 (Continued).

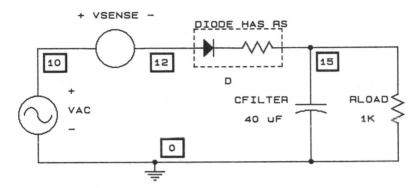

Figure 12.52 Half-Wave Rectifier Circuit.

oratory without some kind of storage device (analog or digital storage oscilloscope).   With SPICE transient analysis this phenomenon can easily be examined in great detail, and insight can be gained into power supply operation.   Figure 12.52 is the schematic diagram of the rectifier circuit, which includes a dead voltage source in series with the diode to measure diode current.

The input file PWRSUPLI.CIR is shown in Fig. 12.53, and shows the diode to be modeled as nearly ideal.   By setting the emission coefficient, N, to 0.001, the forward voltage will be very small compared to a silicon diode.   Refer to Section 11.2 for information on changing diode models.   Voltage source VAC is a 50 Hz sinusoid with an amplitude of 100 Vp.   The voltage across the load resistor and the diode current will be plotted versus time.

Figure 12.54 contains the output file, which plots the load voltage with asterisks and the diode current with plus

```
PWRSUPLI.CIR   1/2 WAVE POWER SUPPLY
VAC  10  0  SIN(0 100  50)
*   VAC IS 100 VOLTS PEAK, 50 HZ
VSENSE 10  12  0
D   12  15  MODA
.MODEL  MODA  D(N = 0.001  RS = 2)
*    DIODE D IS NEAR IDEAL
CFILTER  15  0  40U
RLOAD     15  0  1K
.TRAN 1M  40M
.PLOT TRAN V(15) (0,100) I(VSENSE) (0,1.5)
.OPTIONS NOPAGE
.END
```

Figure 12.53 Input File PWRSUPLI.CIR.

```
********11/22/87******** Demo PSpice (May 1986) *******19:50:49********


PWRSUPLI.CIR   1/2 WAVE POWER SUPPLY


****    CIRCUIT DESCRIPTION


*************************************************************************



VAC  10  0  SIN(0  100   50)
*   VAC IS 100 VOLTS PEAK, 50 HZ
VSENSE  10  12   0
D   12   15   MODA
.MODEL  MODA  D(N = 0.001  RS = 2)
*    DIODE D IS NEAR IDEAL
CFILTER  15   0   40U
RLOAD    15   0   1K
.TRAN 1M  40M
.PLOT TRAN V(15) (0,100) I(VSENSE) (0,1.5)
.OPTIONS NOPAGE
.END

****    DIODE MODEL PARAMETERS



          MODA

IS      1.00D-14

RS        2.000

N          .001

****    INITIAL TRANSIENT SOLUTION    TEMPERATURE =   27.000 DEG C

 NODE   VOLTAGE    NODE   VOLTAGE    NODE   VOLTAGE


( 10)    .0000    ( 12)     .0000    ( 15)     .0000
```

Figure 12.54 Output File PWRSUPLI.OUT.

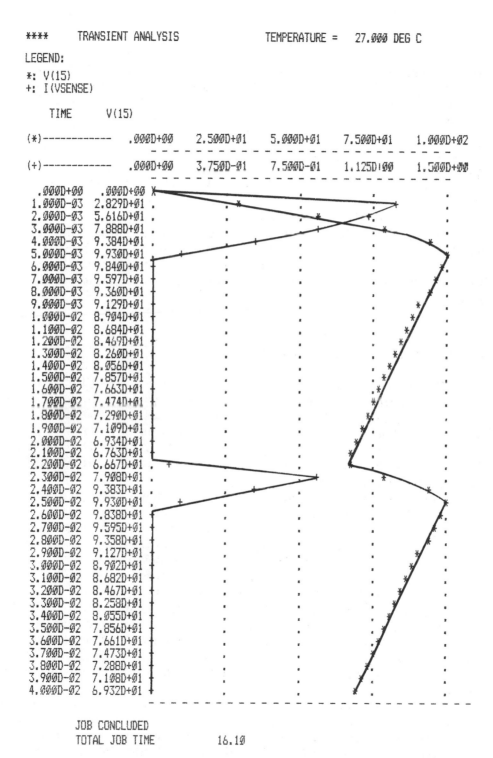

```
****      TRANSIENT ANALYSIS                TEMPERATURE =   27.000 DEG C

LEGEND:

*: V(15)
+: I(VSENSE)

      TIME      V(15)

(*)------------    .000D+00    2.500D+01    5.000D+01    7.500D+01    1.000D+02
                                                        - - - - - - - - - - -
(+)------------    .000D+00    3.750D-01    7.500D-01    1.125D+00    1.500D+00

   .000D+00    .000D+00  X                                               .
  1.000D-03   2.829D+01  .              *                       +        .
  2.000D-03   5.616D+01  .                          *           +        .
  3.000D-03   7.888D+01  .                          +          *         .
  4.000D-03   9.384D+01  .          +                                *   .
  5.000D-03   9.930D+01  +                                             * .
  6.000D-03   9.840D+01  |                                              *
  7.000D-03   9.597D+01  |                                            *  .
  8.000D-03   9.360D+01  |                                        *      .
  9.000D-03   9.129D+01  |                                      *        .
  1.000D-02   8.904D+01  +                                   *           .
  1.100D-02   8.684D+01  |                                 *             .
  1.200D-02   8.469D+01  +                             *                 .
  1.300D-02   8.260D+01  |                           *                   .
  1.400D-02   8.056D+01  |                        *                      .
  1.500D-02   7.857D+01  |                      *                        .
  1.600D-02   7.663D+01  +                    *                          .
  1.700D-02   7.474D+01  |                   *                           .
  1.800D-02   7.290D+01  |                 *                             .
  1.900D-02   7.109D+01  |               *                               .
  2.000D-02   6.934D+01  |             *                                 .
  2.100D-02   6.763D+01  |           *                                   .
  2.200D-02   6.667D+01  +           *                                   .
  2.300D-02   7.908D+01  .                          *                    .
  2.400D-02   9.383D+01  .              +                          *     .
  2.500D-02   9.930D+01  .    +                                        * .
  2.600D-02   9.838D+01  +                                             *.
  2.700D-02   9.595D+01  |                                           *   .
  2.800D-02   9.358D+01  |                                        *      .
  2.900D-02   9.127D+01  |                                     *         .
  3.000D-02   8.902D+01  |                                   *           .
  3.100D-02   8.682D+01  |                                 *             .
  3.200D-02   8.467D+01  +                             *                 .
  3.300D-02   8.258D+01  |                           *                   .
  3.400D-02   8.055D+01  |                        *                      .
  3.500D-02   7.856D+01  |                      *                        .
  3.600D-02   7.661D+01  |                    *                          .
  3.700D-02   7.473D+01  +                   *                           .
  3.800D-02   7.288D+01  |                 *                             .
  3.900D-02   7.108D+01  +               *                               .
  4.000D-02   6.932D+01  +             *                                 .

          JOB CONCLUDED
          TOTAL JOB TIME          16.10
```

Figure 12.54 (Continued).

205

signs. The load voltage rises from 0 V, peaks at 99.3 V, and has the sawtooth shape characteristic of a poorly filtered power supply. The ripple voltage is about 33 Vpp. The diode current during the first positive half cycle (inrush current) peaks at about 1.3 A, while during the second positive half cycle the maximum is around 800 mA. This is because the inrush current must charge the capacitor, which is initially at 0 V, and raise its voltage to 100 V. On subsequent positive half cycles the capacitor voltage never is less than 66 V, so less current is needed to raise its voltage back to 100 V.

To see the diode current more clearly, more plot points would be needed. This could be accomplished by changing the time step in the .TRAN line from 1 ms to a smaller value, such as 0.1 ms. A clearer (but much longer) graph would result.

EXAMPLE 12.5.5, CMOSNAND.CIR

A CMOS (complementary metal-oxide semiconductor) FET inverter NAND circuit is connected to two squarewave voltage sources in this example, and the output is determined as a function of time. The two squarewaves will provide all possible logic input conditions (11, 01, 10, 00 binary) to the NAND gate. The logic NAND gate is made with four enhancement MOSFETs; two are N-channel and two are P-channel.

Figure 12.55 shows the circuit. The NAND gate output is loaded resistively and capacitively.

The input file, Fig. 12.56, includes device capacitances and zero-bias threshold voltages in the .MODEL control lines. For enhancement-mode MOSFETs, VTO is positive for N-channel devices and VTO is negative for P-channel devices. The starting time of the pulses is delayed by 1 us, and the rise and fall times (5 ns) are quite small compared to the pulse widths (5 us and 10 us). Since three plots on one graph gets a little confusing, two graphs are specified. The first will contain the two input voltages, with plot limits that keep the plots separated, and the second is the NAND output voltage by itself.

The graphs in the output file, Fig. 12.57, show that the NAND gate output is low only when both inputs are high, and that the rise and fall times of the output are significant. More resolution of rise and fall times could be obtained by making the time step size smaller in the .TRAN control line.

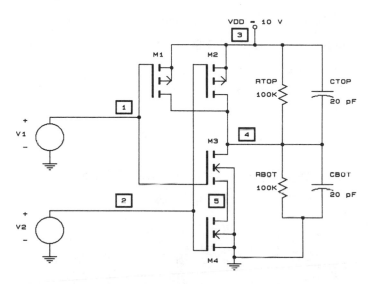

Figure 12.55 CMOS NAND Circuit.

EXAMPLE 12.5.6, TTL.CIR

A 7404 TTL (transistor-transistor logic) inverter logic gate is shown in Fig. 12.58. It is fed by a TTL-compatible squarewave (well, nearly square). The output is connected to a pull-up resistor of 2K ohms.

The input file, Fig. 12.59, describes the input pulse as going from 0 V to 4V after a 10 ns delay, with a 2 ns rise

```
CMOSNAND.CIR    4 TRANSISTOR NAND GATE, CMOS
M1  3  1  4  3  MOD1
M2  3  2  4  3  MOD1
.MODEL MOD1 PMOS(CGDO=10P  CGSO=10P  CBD=.05P  CBS=.05P  VTO=-2)
M3  4  1  5  0  MOD2
M4  5  2  0  0  MOD2
.MODEL MOD2 NMOS(CGDO=10P  CGSO=10P  CBD=.05P  CBS=.05P  VTO=+2)
VDD  3  0  10
RTOP  3  4  100K
CTOP  3  4  20P
RBOT  4  0  100K
CBOT  4  0  20P
V1  1  0  PULSE(0  10  1U  5N  5N  5U  10U)
V2  2  0  PULSE(0  10  1U  5N  5N  10U  20U)
.TRAN 0.8U  24U
.PLOT TRAN V(1) (-15,15)  V(2) (0,30)
.PLOT TRAN V(4) (0,10)
.OPTIONS NOPAGE
.END
```

Figure 12.56 Input File CMOSNAND.CIR.

```
********11/24/87******** Demo PSpice (May 1986) ********10:48:06*********

CMOSNAND.CIR    4 TRANSISTOR NAND GATE, CMOS

****     CIRCUIT DESCRIPTION

***********************************************************************

M1  3  1  4  3  MOD1
M2  3  2  4  3  MOD1
.MODEL MOD1 PMOS(CGDO=10P  CGSO=10P  CBD=.05P  CBS=.05P  VTO=-2)
M3  4  1  5  0  MOD2
M4  5  2  0  0  MOD2
.MODEL MOD2 NMOS(CGDO=10P  CGSO=10P  CBD=.05P  CBS=.05P  VTO=+2)
VDD  3  0  10
RTOP  3  4  100K
CTOP  3  4  20P
RBOT  4  0  100K
CBOT  4  0  20P
V1  1  0  PULSE(0  10  1U  5N  5N  5U  10U)
V2  2  0  PULSE(0  10  1U  5N  5N  10U  20U)
.TRAN 0.8U  24U
.PLOT TRAN V(1) (-15,15)  V(2) (0,30)
.PLOT TRAN V(4) (0,10)
.OPTIONS NOPAGE
.END

****     MOSFET MODEL PARAMETERS


           MOD1      MOD2

TYPE       PMOS      NMOS

LEVEL      1.000     1.000

VTO       -2.000     2.000

KP        2.00D-05  2.00D-05

CBD       5.00D-14  5.00D-14

CBS       5.00D-14  5.00D-14

CGSO      1.00D-11  1.00D-11

CGDO      1.00D-11  1.00D-11

****   INITIAL TRANSIENT SOLUTION     TEMPERATURE =   27.000 DEG C

 NODE   VOLTAGE     NODE   VOLTAGE     NODE   VOLTAGE     NODE   VOLTAGE


 ( 1)    .0000     ( 2)     .0000     ( 3)   10.0000     ( 4)    9.7006

 ( 5)    .0000
```

Figure 12.57 Output File CMOSNAND.OUT.

```
****      TRANSIENT ANALYSIS                 TEMPERATURE =   27.000 DEG C

LEGEND:

*: V(1)
+: V(2)

   TIME      V(1)

(*)-------------- -1.500D+01   -7.500D+00     .000D+00    7.500D+00    1.500D+01
                   - - - - - - - - - - - - - - - - - - - - - - - - - - - -

(+)------------     .000D+00    7.500D+00     1.500D+01    2.250D+01    3.000D+01
                   - - - - - - - - - - - - - - - - - - - - - - - - - - - -
   .000D+00    .000D+00 +          ,              *          ,            ,
 8.000D-07    .000D+00 +          ,              *          ,            ,
 1.600D-06   1.000D+01 .          ,   +          ,          ,   *        ,
 2.400D-06   1.000D+01 .          ,   +          ,          ,   *        ,
 3.200D-06   1.000D+01 .          ,   +          ,          ,   *        ,
 4.000D-06   1.000D+01 .          ,   +          ,          ,   *        ,
 4.800D-06   1.000D+01 .          ,   +          ,          ,   *        ,
 5.600D-06   1.000D+01 .          ,   +          ,          ,   *        ,
 6.400D-06    .000D+00 .          ,   +          X          ,            ,
 7.200D-06    .000D+00 .          ,   +          *          ,            ,
 0.000D-06    .000D+00 .          ,   +          *          ,            ,
 8.800D-06    .000D+00 .          ,   +          *          ,            ,
 9.600D-06    .000D+00 .          ,   +          *          ,            ,
 1.040D-05    .000D+00 .          ,   +          *          ,            ,
 1.120D-05   1.000D+01 +          ,              ,          ,   *        ,
 1.200D-05   1.000D+01 +          ,              ,          ,   *        ,
 1.280D-05   1.000D+01 +          ,              ,          ,   *        ,
 1.360D-05   1.000D+01 +          ,              ,          ,   *        ,
 1.440D-05   1.000D+01 +          ,              ,          ,   *        ,
 1.520D-05   1.000D+01 +          ,              ,          ,   *        ,
 1.600D-05   1.000D+01 +          ,              ,          ,   *        ,
 1.680D-05    .000D+00 +          ,              *          ,            ,
 1.760D-05    .000D+00 +          ,              *          ,            ,
 1.840D-05    .000D+00 +          ,              *          ,            ,
 1.920D-05    .000D+00 +          ,              *          ,            ,
 2.000D-05    .000D+00 +          ,              *          ,            ,
 2.080D-05    .000D+00 +          ,              *          ,            ,
 2.160D-05   1.000D+01 .          ,   +          ,          ,   *        ,
 2.240D-05   1.000D+01 .          ,   +          ,          ,   *        ,
 2.320D-05   1.000D+01 .          ,   +          ,          ,   *        ,
 2.400D-05   1.000D+01 .          ,   +          ,          ,   *        ,
                   - - - - - - - - - - - - - - - - - - - - - - - - - - - -
```

Figure 12.57 (Continued).

```
****      TRANSIENT ANALYSIS                TEMPERATURE =   27.000 DEG C

     TIME       V(4)

                          .000D+00    2.500D+00    5.000D+00    7.500D+00    1.000D+01
               - - - - - - - - - - - - - - - - - - - - - - - - - - - - - - - - - - -
     .000D+00   9.701D+00 .                .            .            .          * .
    8.000D-07   9.701D+00 .                .            .            .          * .
    1.600D-06   4.727D+00 .                .          *.             .            .
    2.400D-06   1.794D+00 .           *    .            .            .            .
    3.200D-06   1.152D+00 .       *        .            .            .            .
    4.000D-06   1.071D+00 .       *        .            .            .            .
    4.800D-06   1.058D+00 .       *        .            .            .            .
    5.600D-06   1.056D+00 .      *         .            .            .            .
    6.400D-06   6.567D+00 .                .            .          *  .           .
    7.200D-06   9.329D+00 .                .            .            .         *  .
    8.000D-06   9.425D+00 .                .            .            .         *  .
    8.800D-06   9.426D+00 .                .            .            .         *  .
    9.600D-06   9.426D+00 .                .            .            .         *  .
    1.040D-05   9.426D+00 .                .            .            .         *  .
    1.120D-05   9.404D+00 .                .            .            .         *  .
    1.200D-05   9.426D+00 .                .            .            .         *  .
    1.280D-05   9.426D+00 .                .            .            .         *  .
    1.360D-05   9.426D+00 .                .            .            .         *  .
    1.440D-05   9.426D+00 .                .            .            .         *  .
    1.520D-05   9.426D+00 .                .            .            .         *  .
    1.600D-05   9.426D+00 .                .            .            .         *  .
    1.680D-05   9.701D+00 .                .            .            .          * .
    1.760D-05   9.701D+00 .                .            .            .          * .
    1.840D-05   9.701D+00 .                .            .            .          * .
    1.920D-05   9.701D+00 .                .            .            .          * .
    2.000D-05   9.701D+00 .                .            .            .          * .
    2.080D-05   9.701D+00 .                .            .            .          * .
    2.160D-05   4.610D+00 .                .          *              .            .
    2.240D-05   1.804D+00 .           *    .            .            .            .
    2.320D-05   1.164D+00 .       *        .            .            .            .
    2.400D-05   1.069D+00 .       *        .            .            .            .
               - - - - - - - - - - - - - - - - - - - - - - - - - - - - - - - - - - -



     JOB CONCLUDED

     TOTAL JOB TIME          78.20
```

Figure 12.57 (Continued).

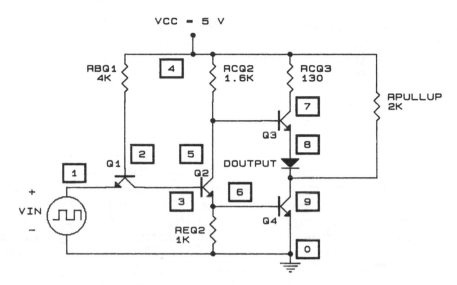

Figure 12.58 TTL Inverter.

time and fall time, pulse width of 20 ns and a period of 48
ns.   The transistor model MOD1 includes a substantial number
of parameters which override the SPICE default parameters for
a BJT.   The optional plot limits cause the two plots to be
separated on the graph, for ease in interpretation.

The output file lists some of the semiconductor parameters, prints an initial transient solution, and has a graph

```
TTL.CIR  TTL 7404 INVERTER GATE, PULSE INPUT
VCC 4 0 5 DC
VIN 1 0 PULSE(0 4 10N 2N 2N 20N 48N)
RBQ1 4 2 4K
RCQ2 4 5 1.6K
REQ2 6 0 1K
RCQ3 4 7 130
RPULLUP 4 9 2K
DOUTPUT 8 9 DMOD1
Q1 3 2 1 MOD1
Q2 5 3 6 MOD1
Q3 7 5 8 MOD1
Q4 9 6 0 MOD1
.WIDTH OUT = 80
.MODEL MOD1 NPN(BF=50 BR=0.1 RB=70 RC=40 TF=.1N TR=10N
+  CJE=0.9P CJC=1.5P CCS=1P VA=50)
.MODEL DMOD1 D
.TRAN 3N 105N
.PLOT TRAN V(9) (0,12) V(1) (-5,5)
.OPTIONS NOPAGE
.END
```

Figure 12.59 Input File TTL.CIR.

of input and output voltages (see Fig. 12.60). The output waveform is valid from a logic standpoint, and shows the distortions that are to be expected from a TTL logic gate operated at 20.8 MHz.

EXAMPLE 12.5.7, ASTABLE.CIR

Transient analyses on astable circuits and free-running oscillators tend to be a bit tricky to accomplish. One must pay careful attention to the initial conditions in the circuit; for example, this circuit (see Fig. 12.61) has complete symmetry (at least on paper). In reality, at the time power is applied to the physical circuit one transistor will go to saturation and the other will become cut-off. As far as SPICE

```
********11/22/87******** Demo PSpice (May 1986) *******19:56:57********


TTL.CIR  TTL 7404 INVERTER GATE, PULSE INPUT


****    CIRCUIT DESCRIPTION


*************************************************************************


VCC 4 0 5 DC
VIN 1 0 PULSE(0 4 10N 2N 2N 20N 48N)
RBQ1 4 2 4K
RCQ2 4 5 1.6K
REQ2 6 0 1K
RCQ3 4 7 130
RPULLUP 4 9 2K
DOUTPUT 8 9 DMOD1
Q1 3 2 1 MOD1
Q2 5 3 6 MOD1
Q3 7 5 8 MOD1
Q4 9 6 0 MOD1
.WIDTH OUT = 80
.MODEL MOD1 NPN(BF=50 BR=0.1 RB=70 RC=40 TF=.1N TR=10N
+  CJE=0.9P CJC=1.5P CCS=1P VA=50)
.MODEL DMOD1 D
.TRAN 3N 105N
.PLOT TRAN V(9) (0,12) V(1) (-5,5)
.OPTIONS NOPAGE
.END
```

Figure 12.60 Output File TTL.OUT.

```
****    DIODE MODEL PARAMETERS


            DMOD1

IS      1.00D-14

****    BJT MODEL PARAMETERS


            MOD1

TYPE       NPN

IS      1.00D-16

BF        50.000

NF         1.000

VAF     5.00D+01

BR          .100

NR         1.000

RB        70.000

RC        40.000

CJE     9.00D-13

TF      1.00D-10

CJC     1.50D-12

TR      1.00D-08

CJS     1.00D-12

****    INITIAL TRANSIENT SOLUTION     TEMPERATURE =   27.000 DEG C

  NODE   VOLTAGE      NODE   VOLTAGE      NODE   VOLTAGE      NODE   VOLTAGE


 (  1)    .0000    (  2)    .8500    (  3)    .0624    (  4)   5.0000

 (  5)   5.0000    (  6)    .0000    (  7)   5.0000    (  8)   5.0000

 (  9)   5.0000
```

Figure 12.60 (Continued).

```
****     TRANSIENT ANALYSIS                  TEMPERATURE =    27.000 DEG C

LEGEND:

*: V(9)
+: V(1)

   TIME       V(9)

(*)-------------     .000D+00     3.000D+00     6.000D+00     9.000D+00     1.200D+01
                   - - - - - - - - - - - - - - - - - - - - - - - - - - - - - - -

(+)-------------  -5.000D+00    -2.500D+00      .000D+00     2.500D+00     5.000D+00
                   - - - - - - - - - - - - - - - - - - - - - - - - - - - - - - -

   .000D+00    5.000D+00 .                 .          *    +              .              .
 3.000D-09    5.000D+00 .                 .          *    +              .              .
 6.000D-09    5.000D+00 .                 .          *    +              .              .
 9.000D-09    5.000D+00 .                 .          *   +               .              .
 1.200D-08    5.224D+00 .                 .          *  ,                .    +         .
 1.500D-08    1.790D+00 .           *      .                .            .    +         .
 1.800D-08    4.324D-01 . *                .                .            .    +         .
 2.100D-08    1.743D-01 .*                 .                .            .    +         .
 2.400D-08    1.771D-01 .*                 .                .            .    +         .
 2.700D-08    1.779D-01 .*                 .                .            .    +         .
 3.000D-08    1.781D-01 .*                 .                .            .    +         .
 3.300D-08    1.762D-01 .*                 .                .        +    .              .
 3.600D-08    1.295D+00 .       *          .            +               .              .
 3.900D-08    2.865D+00 .              *.                +               .              .
 4.200D-08    3.957D+00 .                 .    *         +               .              .
 4.500D-08    4.471D+00 .                 .       *      +               .              .
 4.800D-08    4.741D+00 .                 .          *   +               .              .
 5.100D-08    4.868D+00 .                 .          *   +               .              .
 5.400D-08    4.935D+00 .                 .          *   +               .              .
 5.700D-08    4.968D+00 .                 .          *   +               .              .
 6.000D-08    5.221D+00 .                 .          *  .                .    +         .
 6.300D-08    1.805D+00 .           *      .                .            .    +         .
 6.600D-08    4.505D-01 . *                .                .            .    +         .
 6.900D-08    1.737D-01 .*                 .                .            .    +         .
 7.200D-08    1.773D-01 .*                 .                .            .    +         .
 7.500D-08    1.778D-01 .*                 .                .            .    +         .
 7.800D-08    1.779D-01 .*                 .                .            .    +         .
 8.100D-08    1.764D-01 .*                 .                .        +    .              .
 8.400D-08    1.313D+00 .        *         .            +               .              .
 8.700D-08    2.889D+00 .              *   .            +               .              .
 9.000D-08    3.971D+00 .                 .    *        +               .              .
 9.300D-08    4.476D+00 .                 .       *     +               .              .
 9.600D-08    4.743D+00 .                 .          *  +               .              .
 9.900D-08    4.870D+00 .                 .          *  +               .              .
 1.020D-07    4.935D+00 .                 .          *  +               .              .
 1.050D-07    4.969D+00 .                 .          *  +               .              .
                   - - - - - - - - - - - - - - - - - - - - - - - - - - - - - - -

     JOB CONCLUDED

     TOTAL JOB TIME        105.40
```

Figure 12.60 (Continued).

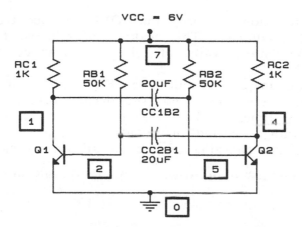

Figure 12.61   BJT Astable Multivibrator.

is concerned, it doesn't know which transistor will win the race to conduct. In order to help SPICE with the transient solution, a good technique is to set some initial conditions which give SPICE a starting point for further analysis. This, of course, requires that you have an understanding of how the circuit starts and runs.

Such initial conditions are specified in the input file in Fig. 12.62, where at time zero Q1 is on and Q2 is off. Of course, in a physical circuit slight differences in transistor parameters and/or stray circuit capacitances might just as likely cause the opposite condition. Notice that the control line .IC . . . has specified five node voltages and that the .TRAN control line includes the UIC (use initial con-

```
ASTABLE.CIR  ASTABLE BJT CIRCUIT
*       OUTPUT IS A SQUAREWAVE
Q1  1  2  0 MOD1
Q2  4  5  0 MOD1
RC1  7  1  1K
RB1  7  2  50K
RC2  7  4  1K
RB2  7  5  50K
CC1B2  1  5 20U
CC2B1  4  2 20U
VCC 7 0 6
.MODEL MOD1 NPN (TF=10N CJC=1P CJE=1P)
.TRAN 0.05 2.5 UIC
.IC V(1)=.1  V(4)=6  V(5)=-6  V(2)=0.8  V(7)=6
.PLOT TRAN V(1)  (0,12) V(2)  (-13,1)
.OPTIONS  NOPAGE  RELTOL=0.01
.END
```

Figure 12.62   Input File ASTABLE.CIR.

ditions) specification.   Also,   the .OPTIONS   control line has
RELTOL=0.01 included.   This makes the relative tolerance 1%
instead of the default value of 0.1%, which saves some computation time.
     The graph in the output file, Fig. 12.63, shows the col-
lector voltage (asterisks) and the base voltage (plus signs)
of Q1 versus time.   The collector voltage is essentially a
squarewave, with a period of about 1.5 seconds, and the base
voltage shows the charging of the 20 uF timing capacitor con-

```
********11/22/87******** Demo PSpice (May 1986) *******17:06:09********


ASTABLE.CIR  ASTABLE BJT CIRCUIT


****     CIRCUIT DESCRIPTION


*****************************************************************************


*      OUTPUT IS A SQUAREWAVE
Q1  1  2  0 MOD1
Q2  4  5  0 MOD1
RC1  7 1  1K
RB1  7 2  50K
RC2  7 4  1K
RB2  7 5  50K
CC1B2  1  5 20U
CC2B1  4  2 20U
VCC 7 0 6
.MODEL MOD1 NPN (TF=10N CJC=1P CJE=1P)
.TRAN 0.05 2.5 UIC
.IC V(1)=.1  V(4)=6  V(5)=-6  V(2)=0.8  V(7)=6
.PLOT TRAN V(1)  (0,12)  V(2)  (-13,1)
.OPTIONS NOPAGE RELTOL=0.01
.END

****     BJT MODEL PARAMETERS



            MOD1

TYPE      NPN

IS       1.00D-16

BF        100.000

NF          1.000

BR          1.000

NR          1.000

CJE      1.00D-12

TF       1.00D-08

CJC      1.00D-12
```

Figure 12.63   Output File ASTABLE.OUT.

```
****     TRANSIENT ANALYSIS              TEMPERATURE =   27.000 DEG C

LEGEND:

*: V(1)
+: V(2)

    TIME      V(1)
(*)------------    .000D+00    3.000D+00    6.000D+00    9.000D+00    1.200D+01
                - - - - - - - - - - - - - - - - - - - -

(+)  ----------- -1.300D+01   -9.500D+00   -6.000D+00   -2.500D+00    1.000D+00

 .000D+00   1.469D-01 .*            .            .            .            +.
 5.000D-02   1.301D-01 .*            .            .            .            +.
 1.000D-01   1.288D-01 .*            .            .            .            +.
 1.500D-01   1.289D-01 .*            .            .            .            +.
 2.000D-01   1.289D-01 .*            .            .            .            +.
 2.500D-01   1.288D-01 .*            .            .            .            +.
 3.000D-01   1.287D-01 .*            .            .            .            +.
 3.500D-01   1.287D-01 .*            .            .            .            +.
 4.000D-01   1.286D-01 .*            .            .            .            +.
 4.500D-01   1.285D-01 .*            .            .            .            +.
 5.000D-01   1.284D-01 .*            .            .            .            +.
 5.500D-01   1.284D-01 .*            .            .            .            +.
 6.000D-01   1.283D-01 .*            .            .            .            +.
 6.500D-01   1.282D-01 .*            .            .            .            +.
 7.000D-01   1.282D-01 .*            .            .            .            +.
 7.500D-01   1.281D-01 .*            .            .            .            +.
 8.000D-01   1.349D-01 .*            .            .            .            +.
 8.500D-01   5.206D+00 .            .            *   .        +            .
 9.000D-01   5.957D+00 .            .            *        +            .
 9.500D-01   6.004D+00 .            .            *            +   .
 1.000D+00   6.000D+00 .            +            *            .    +   .
 1.050D+00   6.000D+00 .            .            *            .  |.
 1.100D+00   6.000D+00 .            .            *            .   .+
 1.150D+00   6.000D+00 .            .            *            .     .+
 1.200D+00   6.000D+00 .            .            *            .       .+
 1.250D+00   6.000D+00 .            .            *            .        +
 1.300D+00   6.000D+00 .            .            *            .         +
 1.350D+00   6.000D+00 .            .            *            .          +
 1.400D+00   6.000D+00 .            .            *            .           +
 1.450D+00   6.000D+00 .            .            *            .           +
 1.500D+00   6.000D+00 .            .            *            .            +.
 1.550D+00   3.499D-02 *            .            .            .            +.
 1.600D+00   8.861D-02 *            .            .            .            +.
 1.650D+00   1.249D-01 .*            .            .            .            +.
 1.700D+00   1.294D-01 .*            .            .            .            +.
 1.750D+00   1.286D-01 .*            .            .            .            +.
 1.800D+00   1.286D-01 .*            .            .            .            +.
 1.850D+00   1.286D-01 .*            .            .            .            +.
 1.900D+00   1.285D-01 .*            .            .            .            +.
 1.950D+00   1.284D-01 .*            .            .            .            +.
 2.000D+00   1.284D-01 .*            .            .            .            +.
 2.050D+00   1.283D-01 .*            .            .            .            +.
 2.100D+00   1.282D-01 .*            .            .            .            +.
 2.150D+00   1.282D-01 .*            .            .            .            +.
 2.200D+00   1.281D-01 .*            .            .            .            +.
 2.250D+00   1.286D-01 .*            .            .            .            +.
 2.300D+00   5.164D+00 .            .            *   .        +            .
 2.350D+00   5.996D+00 .            .            *        +            .
 2.400D+00   6.000D+00 .            .            *        +            .
 2.450D+00   6.000D+00 .            .            *            +   .
 2.500D+00   6.000D+00 .            .            *            +.            .
                - - - - - - - - - - - - - - - - - - - -

    JOB CONCLUDED

    TOTAL JOB TIME       107.20
```

Figure 12.63 (Continued).

nected to it.   The behavior of the circuit once free-running
oscillation has started confirms the validity of the original
initial conditions.

EXAMPLE 12.5.8, ABSVAL.CIR

A real diode has an offset or barrier voltage (about 0.6 V
for silicon) which must be overcome before forward conduction
begins.   This makes real diodes not usable by themselves to
rectify low-level voltages such as audio signals.   Although a
near-ideal diode can be made in a SPICE input file (see Chap.
11.2), in a real circuit another technique must be used to
overcome this limitation.   Figure 12.64 is a precision rectifier circuit, sometimes called an absolute value circuit.
The very large open-loop gain of op-amp X1 is used to negate
the diode offset voltage.   The output voltage of the circuit
is the absolute value of the input voltage; in other words,
it is an ideal full-wave rectifier circuit.
    The input file in Fig. 12.65 models the two operational
amplifiers with subcircuits, in which the subcircuit contains
a large input resistance, R-IN, and a big open-loop voltage
gain, E-BIG.   E-BIG is a VCVS (voltage-controlled voltage
source) with a gain of 50,000 V/V which is controlled by the
differential voltage across nodes 10 and 20 in the subcircuit.
    The diodes are truly ordinary, in that they are modeled
by the SPICE default diode parameters, including an offset
voltage typical of silicon.   The input voltage is a 1 Vp, 100

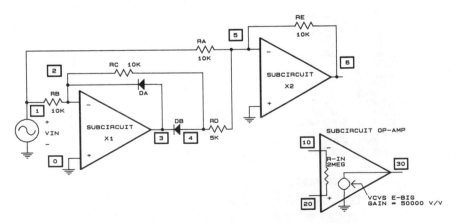

Figure 12.64 Precision Rectifier Circuit.

```
ABSVAL.CIR   ABSOLUTE VALUE (PRECISION RECTIFIER) CIRCUIT, FULL-WAVE
*        2 OP-AMPS IN CIRCUIT, SO SUB-CIRCUIT WILL BE USED FOR OP-AMP
.SUBCKT  OP-AMP  20  10  30
R-IN     20  10  1MEG
E-BIG    30  0   20  10  5E4
*        E-BIG IS A VCVS, GAIN OF 50,000 V/V
.ENDS OP-AMP
RA       1  5  10K
RB       1  2  10K
RC       2  4  10K
RD       4  5  5K
RE       5  6  10K
DA       3  2  ORDINARY
DB       4  3  ORDINARY
.MODEL   ORDINARY D
X1       0  2  3  OP-AMP
X2       0  5  6  OP-AMP
VIN      1  0  SIN(0  1  100)
.TRAN .5M  20M
.PLOT  TRAN  V(6)  (0,2)  V(1)  (-3,1)
.OPTIONS NOPAGE
.END
```

Figure 12.65 Input File ABSVAL.CIR.

Hz sinusoid, and the transient analysis occurs for 2 complete
periods of the input.

In Fig. 12.66, the output file, the input voltage is plotted using plus signs and the output voltage is plotted with
asterisks.   The output voltage is indeed the absolute value
of the input voltage.   The accuracy of this circuit depends
on both the large open-loop voltage gain of the op-amp and
the excellent accuracy of the resistors.

EXAMPLE 12.5.9, AM.CIR

Although SPICE has a built-in model for a frequency modulation generator (SFFM independent source), it does not have an
amplitude modulation generator.   By using a polynomial nonlinear voltage-controlled voltage source (VCVS), an AM generator can easily be created.   The modulation index can changed,
as can the carrier frequency and modulation frequency.   While
the SFFM source is limited to a pure sinusoid as the modulation, the AM generator could be modulated by any kind of waveform.

Figure 12.67 is the schematic diagram of the AM generator.   V-CAR is the carrier sinusoid and V-MOD is the modulation voltage, which happens to be a sinusoid with a DC offset
voltage.   It could be a pulse, exponential, piece-wise linear

```
********11/20/87******** Demo PSpice (May 1986) *******15:23:09********

ABSVAL.CIR    ABSOLUTE VALUE (PRECISION RECTIFIER) CIRCUIT, FULL-WAVE

****     CIRCUIT DESCRIPTION

**************************************************************************


*      2 OP-AMPS IN CIRCUIT, SO SUB-CIRCUIT WILL BE USED FOR OP-AMP
.SUBCKT  OP-AMP  20  10  30
R-IN   20  10  1MEG
E-BIG  30  0  20  10  5E4
*      E-BIG IS A VCVS, GAIN OF 50,000 V/V
.ENDS OP-AMP
RA     1  5  10K
RB     1  2  10K
RC     2  4  10K
RD     4  5  5K
RE     5  6  10K
DA     3  2  ORDINARY
DB     4  3  ORDINARY
.MODEL  ORDINARY D
X1     0  2  3  OP-AMP
X2     0  5  6  OP-AMP
VIN    1  0  SIN(0  1  100)
.TRAN  .5M  20M
.PLOT  TRAN  V(6)  (0,2)  V(1)  (-3,1)
.OPTIONS NOPAGE
.END

****     DIODE MODEL PARAMETERS



         ORDINARY

IS       1.00D-14

****     INITIAL TRANSIENT SOLUTION      TEMPERATURE =   27.000 DEG C

 NODE  VOLTAGE      NODE  VOLTAGE      NODE  VOLTAGE      NODE  VOLTAGE


(  1)    .0000   (  2)    .0000   (  3)    .0000   (  4)     .0000

(  5)    .0000   (  6)    .0000
```

Figure 12.66 Output File ABSVAL.OUT.

```
****        TRANSIENT ANALYSIS                TEMPERATURE =   27.000 DEG C
LEGEND:
*: V(6)
+: V(1)
      TIME       V(6)
   (*)-------------    .000D+00    5.000D-01    1.000D+00    1.500D+00    2.000D+00
                    - - - - - - - - - - - - - - - - - - - - - - - - - - - - - - -
   (+)---------  - -3.000D+00   -2.000D+00   -1.000D+00     .000D+00    1.000D+00
                    - - - - - - - - - - - - - - - - - - - - - - - - - - - - - - -
    .000D+00    .000D+00 *        .         .            .         +         .
   5.000D-04   3.088D-01 .        *         .            .            +      .
   1.000D-03   5.842D-01 .        .  *      .            .              +    .
   1.500D-03   8.025D-01 .        .         *            .               +   .
   2.000D-03   9.449D-01 .        .         .    *.      .                +  .
   2.500D-03   9.989D-01 .        .         .       *    .                  +
   3.000D-03   9.458D-01 .        .         .    *.      .                +  .
   3.500D-03   8.025D-01 .        .         *            .               +   .
   4.000D-03   5.837D-01 .        .  *      .            .              +    .
   4.500D-03   3.086D-01 .        *         .            .            +      .
   5.000D-03   8.627D-02 . *      .         .            .         +         .
   5.500D-03   3.065D-01 .        *         .            .      +            .
   6.000D-03   5.842D-01 .        .  *      .         +  .                   .
   6.500D-03   8.083D-01 .        .         *   . +     .                   .
   7.000D-03   9.458D-01 .        .         . *.+       .                   .
   7.500D-03   9.921D-01 .        .         .    X      .                   .
   8.000D-03   9.449D-01 .        .         . *.+       .                   .
   8.500D-03   8.082D-01 .        .         *   . +     .                   .
   9.000D-03   5.848D-01 .        .  *      .         +  .                   .
   9.500D-03   3.066D-01 .        *         .            .      +            .
   1.000D-02   1.007D-01 .   *    .         .            .         +         .
   1.050D-02   3.087D-01 .        *         .            .            +      .
   1.100D-02   5.842D-01 .        .  *      .            .              +    .
   1.150D-02   8.025D-01 .        .         *            .               +   .
   1.200D-02   9.449D-01 .        .         .    *.      .                +  .
   1.250D-02   9.989D-01 .        .         .       *    .                  +
   1.300D-02   9.458D-01 .        .         .    *.      .                +  .
   1.350D-02   8.025D-01 .        .         *            .               +   .
   1.400D-02   5.837D-01 .        .  *      .            .              +    .
   1.450D-02   3.086D-01 .        *         .            .            +      .
   1.500D-02   8.627D-02 . *      .         .            .         +         .
   1.550D-02   3.065D-01 .        *         .            .      +            .
   1.600D-02   5.842D-01 .        .  *      .         +  .                   .
   1.650D-02   8.083D-01 .        .         *   . +     .                   .
   1.700D-02   9.458D-01 .        .         . *.+       .                   .
   1.750D-02   9.921D-01 .        .         .    X      .                   .
   1.800D-02   9.449D-01 .        .         . *.+       .                   .
   1.850D-02   8.082D-01 .        .         *   . +     .                   .
   1.900D-02   5.848D-01 .        .  *      .         +  .                   .
   1.950D-02   3.066D-01 .        *         .            .      +            .
   2.000D-02   1.061D-14 *        .         .            .         +         .
                    - - - - - - - - - - - - - - - - - - - - - - - - - - - - - - -
            JOB CONCLUDED
            TOTAL JOB TIME        27.40
```

Figure 12.66 (Continued).

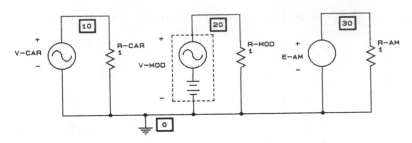

Figure 12.67   Amplitude Modulation Generator.

or even SFFM source. Both of these generators will be used to control the output of the VCVS E-AM, across which the AM carrier will appear. Since SPICE requires that each node have more than one element connected to it, three dummy resistors (R-CAR, R-MOD and R-AM) are used to satisfy this requirement.

The input file in Fig. 12.68 specifies the carrier voltage to be a 10 Vp 10 KHz sinusoid and the modulation voltage to be 1 VDC in series with a 1 Vp, 2 KHz sinusoid which is delayed 200 us from time zero. The polynomial VCVS E-AM multiplies V-CAR by V-MOD because the p4 coefficient (see Appendix C.1.2) is 1 and all other coefficients are 0. The equation for E-AM would then be

$$e(t) = (1 + 1 \sin(2*PI*2K*t)) \ (10 \sin(2*PI*10K*t))$$

which is the equation of a 100% modulated AM, or DSB-FC, carrier. The % modulation can be decreased by lowering the amplitude of the 2 KHz sinusoid from 1 Vp. The only change need-

```
AM.CIR   AMPLITUDE MODULATION USING A POLYNOMIAL VCVS
*        POLYNOMIAL CONTROLLED SOURCES ARE COVERED IN APPENDIX C
V-CAR    10  0  SIN(0  10  10K)
*        V-CAR IS 10 VP 10 KHZ SINE, CARRIER
R-CAR    10  0  1
V-MOD    20  0  SIN(1  1  2K  200U)
*        V-MOD IS 1VDC + 1 VP 2 KHZ SINE, DELAYED 200 US, MODULATION
R-MOD    20  0  1
E-AM     30  0  POLY(2)  10  0  20  0  0  0  0  0  0  1
R-AM     30  0  1
.TRAN 10U  1M
.OPTIONS  NOPAGE
.PLOT TRAN  V(30)
.END
```

Figure 12.68   Input File AM.CIR.

ed to change E-AM into a double sideband suppressed carrier is to change the DC offset voltage from 1 V to 0 V.

Figure 12.69 contains the output file which shows an unmodulated carrier for the first 200 us, at which time the modulation envelope of a 2 KHz sinusoid begins. The carrier reaches its maximum values at 325 us and 825 us; the minimum occurs at 575 us when the carrier amplitude drops to 0 V.

12.6 .TEMP, .FOUR, .TF, .OP, .DISTO, .SENS & .NOISE CONTROL LINES

EXAMPLE 12.6.1, VARYTEMP.CIR

In order to examine the sensitivity of diode current to temperature, a diode with SPICE default parameters is connected to a DC voltage source which can be stepped through a range using a DC analysis. The DC analysis results can be used to plot a volt-ampere characteristic of the diode. Figure 12.70 shows the dead voltage source VSENSE connected in series to measure diode current.

```
********11/22/87******** Demo PSpice (May 1986) *******19:31:34********


AM.CIR   AMPLITUDE MODULATION USING A POLYNOMIAL VCVS

****     CIRCUIT DESCRIPTION

************************************************************************

*     POLYNOMIAL CONTROLLED SOURCES ARE COVERED IN APPENDIX C
V-CAR 10 0 SIN(0 10 10K)
*     V-CAR IS 10 VP 10 KHZ SINE, CARRIER
R-CAR 10 0 1
V-MOD       20 0 SIN(1 1 2K 200U)
*     V-MOD IS 1VDC + 1 VP 2 KHZ SINE, DELAYED 200 US, MODULATION
R-MOD 20 0 1
E-AM  30 0 POLY(2) 10 0 20 0 0 0 0 0 1
R-AM  30 0 1
.TRAN 10U 1M
.OPTIONS NOPAGE
.PLOT TRAN V(30)
.END

****     INITIAL TRANSIENT SOLUTION     TEMPERATURE =   27.000 DEG C

  NODE   VOLTAGE     NODE   VOLTAGE     NODE   VOLTAGE
 ( 10)    .0000    ( 20)   1.0000    ( 30)    .0000
```

Figure 12.69 Output File AM.OUT.

```
      TIME      V(30)

                  -2.000D+01   -1.000D+01     .000D+00    1.000D+01   2.000D+01
             - - - - - - - - - - - - - - - - - - - - - - - - - - - - - - - - -
  .000D+00    .000D+00  .              .             *            ,           .
 1.000D-05   5.764D+00  .              .             .         *  ,           .
 2.000D-05   9.148D+00  .              .             .            *.          .
 3.000D-05   9.126D+00  .              .             .            *.          .
 4.000D-05   5.618D+00  .              .             .         *  .           .
 5.000D-05  -3.624D-02  .              .             *            .           .
 6.000D-05  -5.676D+00  .              .      *      ,            .           .
 7.000D-05  -9.148D+00  .              .*            .            .           .
 8.000D-05  -9.126D+00  .              .*            .            .           .
 9.000D-05  -5.618D+00  .              .      *      .            .           .
 1.000D-04   3.624D-02  .              .             *            .           .
 1.100D-04   5.676D+00  .              .             .         *  .           .
 1.200D-04   9.148D+00  .              .             ,            *.          .
 1.300D-04   9.126D+00  .              .             .            *.          .
 1.400D-04   5.618D+00  .              .             ,         *  .           .
 1.500D-04  -3.625D-02  .              .             *            .           .
 1.600D-04  -5.676D+00  .              .      *      ,            .           .
 1.700D-04  -9.148D+00  .              .*            .            .           .
 1.800D-04  -9.126D+00  .              .*            .            .           .
 1.900D-04  -5.618D+00  .              .      *      ,            .           .
 2.000D-04  -7.595D-14  .              ,             *            ,           .
 2.100D-04   6.497D+00  .              ,             .          * .           .
 2.200D-04   1.135D+01  .              .             ,            . *         .
 2.300D-04   1.231D+01  .              .             .            .   *       .
 2.400D-04   8.140D+00  .              .             ,            *.          .
 2.500D-04  -1.570D-01  .              .             *            .           .
 2.600D-04  -9.524D+00  .              .*            ,            .           .
 2.700D-04  -1.604D+01  .      *       .             .            ,           .
 2.800D-04  -1.663D+01  .    *         .             .            .           .
 2.900D-04  -1.057D+01  .         *.   .             .            .           .
 3.000D-04   5.998D-02  .              .             *            .           .
 3.100D-04   1.110D+01  .              .             .          .*            .
 3.200D-04   1.806D+01  .              .             .            ,           *
 3.300D-04   1.806D+01  .              .             .            ,           *
 3.400D-04   1.110D+01  .              .             ,            .*           .
 3.500D-04   5.998D-02  .              .             *            .           .
 3.600D-04  -1.057D+01  .         *.   .             .            .           .
 3.700D-04  -1.663D+01  .      *       .             .            .           .
 3.800D-04  -1.604D+01  .       *      .             .            .           .
 3.900D-04  -9.524D+00  .              .*            .            .           .
 4.000D-04  -1.570D-01  .              .             *            ,           .
 4.100D-04   8.140D+00  .              .             .         *  .           .
 4.200D-04   1.231D+01  .              .             .            , *         .
 4.300D-04   1.135D+01  .              .             .            . *         .
 4.400D-04   6.445D+00  .              .             ,         *  .           .
 4.500D-04   1.940D-01  .              .             *            .           .
 4.600D-04  -4.735D+00  .              .          *  ,            .           .
 4.700D-04  -6.742D+00  .              .     *       ,            .           .
 4.800D-04  -5.778D+00  .              .       *     .            .           .
 4.900D-04  -3.040D+00  .              .           * .            .           .
```

Figure 12.69 (Continued).

```
                 -2.000D+01   -1.000D+01     .000D+00    1.000D+01    2.000D+01
5.000D-04 -1.570D-01 .          .            *           .            .
5.100D-04  1.657D+00 .          .            . *         .            .
5.200D-04  2.051D+00 .          .            . *         .            .
5.300D-04  1.455D+00 .          .            . *         .            .
5.400D-04  6.089D-01 .          .            .*          .            .
5.500D-04  5.996D-02 .          .            *           .            .
5.600D-04 -8.044D-02 .          .            *           .            .
5.700D-04 -3.190D-02 .          .            *           .            .
5.800D-04 -3.190D-02 .          .            *           .            .
5.900D-04 -8.044D-02 .          .            *           .            .
6.000D-04  5.996D-02 .          .            *           .            .
6.100D-04  6.088D-01 .          .            .*          .            .
6.200D-04  1.455D+00 .          .            . *         .            .
6.300D-04  2.051D+00 .          .            . *         .            .
6.400D-04  1.657D+00 .          .            . *         .            .
6.500D-04 -1.570D-01 .          .            *           .            .
6.600D-04 -3.040D+00 .          .        *   .           .            .
6.700D-04 -5.778D+00 .          .    *       .           .            .
6.800D-04 -6.742D+00 .          .   *        .           .            .
6.900D-04 -4.735D+00 .          .     *      .           .            .
7.000D-04  1.940D-01 .          .            *           .            .
7.100D-04  6.445D+00 .          .            .       *   .            .
7.200D-04  1.135D+01 .          .            .           .*           .
7.300D-04  1.231D+01 .          .            .           . *          .
7.400D-04  8.140D+00 .          .            .         * .            .
7.500D-04 -1.570D-01 .          .            *           .            .
7.600D-04 -9.524D+00 .          . *          .           .            .
7.700D-04 -1.604D+01 .      *    .            .           .            .
7.800D-04 -1.663D+01 .    *      .            .           .            .
7.900D-04 -1.057D+01 .        *. .            .           .            .
8.000D-04  5.992D-02 .          .            *           .            .
8.100D-04  1.110D+01 .          .            .          .*            .
8.200D-04  1.806D+01 .          .            .           .          * .
8.300D-04  1.806D+01 .          .            .           .          * .
8.400D-04  1.110D+01 .          .            .          .*            .
8.500D-04  5.992D-02 .          .            *           .            .
8.600D-04 -1.057D+01 .        *. .            .           .            .
8.700D-04 -1.663D+01 .    *      .            .           .            .
8.800D-04 -1.604D+01 .     *     .            .           .            .
8.900D-04 -9.524D+00 .          .*           .           .            .
9.000D-04 -1.570D-01 .          .            *           .            .
9.100D-04  8.140D+00 .          .            .         * .            .
9.200D-04  1.231D+01 .          .            .           . *          .
9.300D-04  1.135D+01 .          .            .           .*           .
9.400D-04  6.445D+00 .          .            .       *   .            .
9.500D-04  1.940D-01 .          .            *           .            .
9.600D-04 -4.735D+00 .          .     *      .           .            .
9.700D-04 -6.742D+00 .          .   *        .           .            .
9.800D-04 -5.778D+00 .          .    *       .           .            .
9.900D-04 -3.040D+00 .          .        *   .           .            .
1.000D-03  1.919D-14 .          .            *           .            .

      JOB CONCLUDED

      TOTAL JOB TIME        21.70
```

Figure 12.69 (Continued).

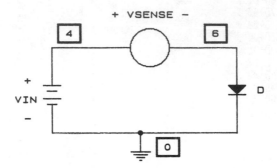

Figure 12.70 Diode Test Circuit.

The input file, Fig. 12.71, shows that the DC source is stepped from 0.65 V to 0.80 V. The .TEMP control line specifies that the DC analysis will occur at a circuit temperature of 50 deg C, and will be repeated at 100 deg C. For each analysis, a plot will be made of diode current versus diode voltage.

The graphs in the output file, Fig. 12.72, look very similar at first glance until you realize that SPICE has scaled them differently. For example, for a diode forward voltage of 0.7 V the diode current is 21.88 mA at 50 deg C and 242.3 mA at 100 deg C. The effect of temperature on diode current could be made clearer by using plot limits to make the current scales the same on both graphs.

EXAMPLE 12.6.2, DELTAT.CIR

The effect of temperature on a BJT differential amplifier with a constant current source is done using the .TEMP control line. Figure 12.73 shows the circuit, which includes a

```
VARYTEMP.CIR    PLOTS DIODE I-V CURVES AT DIFF TEMPS.
VIN 4 0 DC 0.5
D   6 0 TYPE1
VSENSE  4  6  0
.MODEL TYPE1 D
.DC VIN .65  .80  .005
.PLOT  DC  I(VSENSE)
.OPTIONS NOPAGE
.TEMP 50  100
.END
```

Figure 12.71 Input File VARYTEMP.CIR.

single-ended input voltage source VIN connected to the base of Q1.

The input file, Fig. 12.74, includes a .TF (transfer function) control line. This is explored in Ex. 12.6.4 as well. The transfer function will give the input resistance at VIN, the output resistance at node 3, and the voltage gain from VIN to the collector of Q1 (node 3).

Output file DELTAT.OUT (Fig. 12.75) includes a complete analysis of the BJT model parameters, a small-signal bias solution, operating point information about each transistor and small-signal characteristics (the result of the .TF line) at both temperatures specified in the .TEMP line.

At 0 deg C, the voltage gain is -71.85 V/V while at 100 deg C, the voltage gain has decreased to -51.71 V/V.

```
********11/22/87******** Demo PSpice (May 1986) *******20:01:47********


VARYTEMP.CIR    PLOTS DIODE I-V CURVES AT DIFF TEMPS.


****    CIRCUIT DESCRIPTION


*************************************************************************


VIN 4 0 DC 0.5
D    6 0  TYPE1
VSENSE  4  6  0
.MODEL TYPE1 D
.DC VIN .65  .80  .005
.PLOT  DC  I(VSENSE)
.OPTIONS NOPAGE
.TEMP 50  100
.END


****    DIODE MODEL PARAMETERS


          TYPE1

IS      1.00D-14
```

Figure 12.72 Output File VARYTEMP.OUT.

```
****    TEMPERATURE-ADJUSTED VALUES       TEMPERATURE =   50.000 DEG C



**** DIODE MODEL PARAMETERS

NAME       IS        VJ       CJO


TYPE1    2.647D-13  9.787D-01    .000D+00

****    DC TRANSFER CURVES              TEMPERATURE =   50.000 DEG C


  VIN        I(VSENSE)

                  .000D+00    2.000D-01    4.000D-01    6.000D-01    8.000D-01
             - - - - - - - - - - - - - - - - - - - - - - - - - - - - - - -
 6.500D-01  3.633D-03 *          .            .            .            .
 6.550D-01  4.348D-03 *          .            .            .            .
 6.600D-01  5.203D-03 *          .            .            .            .
 6.650D-01  6.227D-03 *          .            .            .            .
 6.700D-01  7.452D-03 *          .            .            .            .
 6.750D-01  8.917D-03 .*         .            .            .            .
 6.800D-01  1.067D-02 .*         .            .            .            .
 6.850D-01  1.277D-02 .*         .            .            .            .
 6.900D-01  1.528D-02 .*         .            .            .            .
 6.950D-01  1.829D-02 .*         .            .            .            .
 7.000D-01  2.188D-02 .*         .            .            .            .
 7.050D-01  2.619D-02 . *        .            .            .            .
 7.100D-01  3.134D-02 . *        .            .            .            .
 7.150D-01  3.750D-02 . *        .            .            .            .
 7.200D-01  4.488D-02 .  *       .            .            .            .
 7.250D-01  5.371D-02 .  *       .            .            .            .
 7.300D-01  6.427D-02 .   *      .            .            .            .
 7.350D-01  7.691D-02 .     *    .            .            .            .
 7.400D-01  9.204D-02 .      *   .            .            .            .
 7.450D-01  1.101D-01 .       *  .            .            .            .
 7.500D-01  1.318D-01 .         * .           .            .            .
 7.550D-01  1.577D-01 .          * .          .            .            .
 7.600D-01  1.888D-01 .            *.         .            .            .
 7.650D-01  2.259D-01 .            . *        .            .            .
 7.700D-01  2.703D-01 .            .    *     .            .            .
 7.750D-01  3.235D-01 .            .       *  .            .            .
 7.800D-01  3.871D-01 .            .          *.           .            .
 7.850D-01  4.632D-01 .            .           .    *      .            .
 7.900D-01  5.544D-01 .            .           .          *.            .
 7.950D-01  6.634D-01 .            .           .           .     *      .
 8.000D-01  7.939D-01 .            .           .           .            *
             - - - - - - - - - - - - - - - - - - - - - - - - - - - - - - -
```

Figure 12.72 (Continued).

```
****    TEMPERATURE-ADJUSTED VALUES     TEMPERATURE =  100.000 DEG C



****  DIODE MODEL PARAMETERS

NAME         IS         VJ        CJO


TYPE1     8.509D-11  9.299D-01    .000D+00

****    DC TRANSFER CURVES           TEMPERATURE =  100.000 DEG C


   VIN        I(VSENSE)

               .000D+00     2.000D+00     4.000D+00     6.000D+00     8.000D+00
             ---------------------------------------------------------------------
 6.500D-01  5.118D-02 *              .             .             .             .
 6.550D-01  5.979D-02 *              .             .             .             .
 6.600D-01  6.985D-02 *              .             .             .             .
 6.650D-01  8.160D-02 .*             .             .             .             .
 6.700D-01  9.532D-02 .*             .             .             .             .
 6.750D-01  1.114D-01 .*             .             .             .             .
 6.800D-01  1.301D-01 .*             .             .             .             .
 6.850D-01  1.520D-01 .*             .             .             .             .
 6.900D-01  1.776D-01 .*             .             .             .             .
 6.950D-01  2.074D-01 .*             .             .             .             .
 7.000D-01  2.423D-01 . *            .             .             .             .
 7.050D-01  2.831D-01 . *            .             .             .             .
 7.100D-01  3.307D-01 . *            .             .             .             .
 7.150D-01  3.864D-01 .  *           .             .             .             .
 7.200D-01  4.514D-01 .  *           .             .             .             .
 7.250D-01  5.273D-01 .  *           .             .             .             .
 7.300D-01  6.160D-01 .    *         .             .             .             .
 7.350D-01  7.196D-01 .     *        .             .             .             .
 7.400D-01  8.407D-01 .     *        .             .             .             .
 7.450D-01  9.822D-01 .      *       .             .             .             .
 7.500D-01  1.147D+00 .        *     .             .             .             .
 7.550D-01  1.340D+00 .         *   .              .             .             .
 7.600D-01  1.566D+00 .          * ,               .             .             .
 7.650D-01  1.829D+00 .           *,               .             .             .
 7.700D-01  2.137D+00 .             .*             .             .             .
 7.750D-01  2.497D+00 .             .  *           .             .             .
 7.800D-01  2.917D+00 .             .      *       .             .             .
 7.850D-01  3.408D+00 .             .         *    .             .             .
 7.900D-01  3.981D+00 .             .             *             .             .
 7.950D-01  4.651D+00 .             .             .  *          .             .
 8.000D-01  5.433D+00 .             .             .       *     .             .
             ---------------------------------------------------------------------

        JOB CONCLUDED

        TOTAL JOB TIME          9.40
```

Figure 12.72 (Continued).

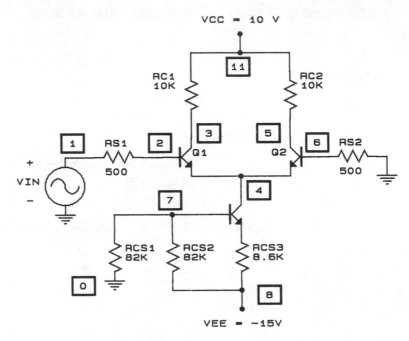

Figure 12.73 Differential Amplifier Circuit.

```
DELTAT.CIR    DIFF. AMP. OP. POINT AT 3 TEMPERATURES
RS1   1   2   500
RS2   6   0   500
VIN   1   0   AC
RC1   11  3   10K
RC2   11  5   10K
VCC   11  0   10
VEE   9   0   -15
Q1    3   2   4   MOD1
Q2    5   6   4   MOD1
QCS   4   7   8   MOD1
RCS1  7   0   82K
RCS2  7   9   82K
RCS3  8   9   8.6K
.MODEL MOD1 NPN
.TF  V(3) VIN
.TEMP 0 125
.OPTIONS NOPAGE
.END
```

Figure 12.74 Input File DELTAT.CIR.

```
********11/22/87******** Demo PSpice (May 1986) *******19:36:59********

DELTAT.CIR   DIFF. AMP. OP. POINT AT 3 TEMPERATURES

****    CIRCUIT DESCRIPTION


************************************************************************


RS1  1  2  500
RS2  6  0  500
VIN  1  0  AC
RC1  11  3  10K
RC2  11  5  10K
VCC  11  0  10
VEE  9  0  -15
Q1  3  2  4  MOD1
Q2  5  6  4  MOD1
QCS  4  7  8  MOD1
RCS1  7  0  82K
RCS2  7  9  82K
RCS3  8  9  8.6K
.MODEL MOD1 NPN
.TF  V(3) VIN
.TEMP  0  125
.OPTIONS NOPAGE
.END

****    BJT MODEL PARAMETERS



          MOD1

TYPE      NPN

IS      1.00D-16

BF       100.000

NF         1.000

BR         1.000

NR         1.000
```

Figure 12.75 Output File DELTAT.OUT.

```
****      TEMPERATURE-ADJUSTED VALUES      TEMPERATURE =      .000 DEG C

**** BJT MODEL PARAMETERS

NAME        IS         BF         ISE        BR         ISC        VJE        VJC

MOD1     1.084D-18 1.000D+02  .000D+00 1.000D+00  .000D+00 7.965D-01 7.965D-01

****      SMALL SIGNAL BIAS SOLUTION        TEMPERATURE =      .000 DEG C

  NODE   VOLTAGE      NODE   VOLTAGE      NODE   VOLTAGE      NODE   VOLTAGE

 (  1)    .0000     (  2)    -.0018     (  3)    6.3556     (  4)    -.7891

 (  5)   6.3556     (  6)    -.0018     (  7)   -7.8018     (  8)   -8.6057

 (  9)  -15.0000    ( 11)   10.0000


    VOLTAGE SOURCE CURRENTS

    NAME       CURRENT


    VIN       -3.644D-06

    VCC       -7.289D-04

    VEE        8.313D-04


    TOTAL POWER DISSIPATION   1.98D-02  WATTS

****      OPERATING POINT INFORMATION      TEMPERATURE =      .000 DEG C

**** BIPOLAR JUNCTION TRANSISTORS

            Q1        Q2        QCS

MODEL     MOD1      MOD1      MOD1
IB        3.64E-06  3.64E-06  7.36E-06
IC        3.64E-04  3.64E-04  7.36E-04
VBE        .787      .787      .804
VBC      -6.36     -6.36     -7.01
VCE       7.14      7.14      7.82
BETADC  100.      100.      100.
GM        1.55E-02  1.55E-02  3.13E-02
RPI       6.46E+03  6.46E+03  3.20E+03
RX         .00E+00   .00E+00   .00E+00
RO        1.00E+12  1.00E+12  1.00E+12
CPI        .00E+00   .00E+00   .00E+00
CMU        .00E+00   .00E+00   .00E+00
CBX        .00E+00   .00E+00   .00E+00
CCS        .00E+00   .00E+00   .00E+00
BETAAC  100.      100.      100.
FT        2.46E+17  2.46E+17  4.98E+17
```

Figure 12.75 (Continued).

```
****      SMALL-SIGNAL CHARACTERISTICS

     V(3)/VIN                          =  -7.185D+01

     INPUT RESISTANCE AT VIN           =   1.392D+04

     OUTPUT RESISTANCE AT V(3)         =   1.000D+04



****      TEMPERATURE-ADJUSTED VALUES      TEMPERATURE =  125.000 DEG C



**** BJT MODEL PARAMETERS

NAME        IS        BF        ISE       BR        ISC       VJE       VJC

MOD1     9.031D-12 1.000D+02  .000D+00 1.000D+00  .000D+00 5.727D-01 5.727D-01

****      SMALL SIGNAL BIAS SOLUTION      TEMPERATURE =  125.000 DEG C


  NODE    VOLTAGE     NODE    VOLTAGE     NODE    VOLTAGE     NODE    VOLTAGE


 (  1)     .0000    (  2)    -.0019    (  3)     6.2588    (  4)    -.6036

 (  5)    6.2588    (  6)    -.0019    (  7)    -7.8098    (  8)    -8.4357

 (  9)  -15.0000    ( 11)   10.0000
```

```
    VOLTAGE SOURCE CURRENTS

    NAME       CURRENT

    VIN      -3.741D-06

    VCC      -7.482D-04

    VEE       8.510D-04

    TOTAL POWER DISSIPATION   2.02D-02  WATTS
```

Figure 12.75 (Continued).

```
****      OPERATING POINT INFORMATION      TEMPERATURE =  125.000 DEG C


**** BIPOLAR JUNCTION TRANSISTORS


                Q1          Q2          QCS

MODEL    MOD1        MOD1        MOD1
IB       3.74E-06    3.74E-06    7.56E-06
IC       3.74E-04    3.74E-04    7.56E-04
VBE      .602        .602        .626
VBC      -6.26       -6.26       -7.21
VCE      6.86        6.86        7.83
BETADC   100.        100.        100.
GM       1.09E-02    1.09E-02    2.20E-02
RPI      9.17E+03    9.17E+03    4.54E+03
RX       .00E+00     .00E+00     .00E+00
RO       1.00E+12    1.00E+12    1.00E+12
CPI      .00E+00     .00E+00     .00E+00
CMU      .00E+00     .00E+00     .00E+00
CBX      .00E+00     .00E+00     .00E+00
CCS      .00E+00     .00E+00     .00E+00
BETAAC   100.        100.        100.
FT       1.74E+17    1.74E+17    3.51E+17



****      SMALL-SIGNAL CHARACTERISTICS

    V(3)/VIN                           = -5.171D+01

    INPUT RESISTANCE AT VIN            =  1.934D+04

    OUTPUT RESISTANCE AT V(3)          =  1.000D+04


        JOB CONCLUDED

        TOTAL JOB TIME          8.80
```

Figure 12.75 (Continued).

EXAMPLE 12.6.3, FOURIER.CIR

A Fourier analysis of a waveform can be done by SPICE in conjunction with a transient analysis.   The circuit of Fig.
12.76 is a half-wave rectifier with a resistive load.   If the
correct fundamental frequency can be specified in the .FOUR
control line, the amplitude and phase of the first nine harmonics present in the waveform will be calculated and printed
along with the DC value of the waveform.   The fundamental frequency in this case should be the input sinusoid frequency.
    The input file, Fig. 12.77, shows the input source fre-
quency to be 100 Hz, with a period of 10 ms and a peak voltage of 1 Vp.   The transient analysis is done for 10 ms, and
the fundamental frequency is clearly 100 Hz.   The diode model
is modified by setting N to .001 to make the diode nearly
ideal.

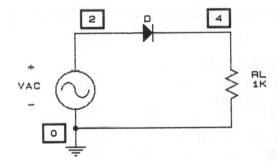

Figure 12.76   Half-wave Rectifier Circuit.

```
FOURIER.CIR     1/2-WAVE RECTIFIER W/FOURIER ANALYSIS
VAC  2  0  SIN(0  1.0  100)
D    2  4  IDEAL
RL   4  0  1K
.MODEL   IDEAL  D(N=0.001)
.TRAN .25M 10M
.FOUR  100  V(4,0)
.PLOT  TRAN V(4,0)  V(2,0)
.OPTIONS NOPAGE
.END
```

Figure 12.77   Input File FOURIER.CIR.

   The graph in the output file (Fig. 12.78) has two plots:
the input sinusoid (plus signs) and the rectified output voltage (asterisks).   At the peak of the output voltage the amplitude is 0.998 V, indicating that the diode is dropping only 2
mV.
   The Fourier component table lists the DC value to be
0.3178 V, very close to the theoretical value of 1/PI V.   The
amplitudes and phases of the first nine harmonics are listed
as well.

```
********11/22/87******** Demo PSpice (May 1986) *******19:44:19********


FOURIER.CIR      1/2-WAVE RECTIFIER W/FOURIER ANALYSIS


****     CIRCUIT DESCRIPTION


***************************************************************************


VAC  2  0  SIN(0  1.0  100)
D    2  4  IDEAL
RL   4  0  1K
.MODEL  IDEAL D(N=0.001)
.TRAN .25M 10M
.FOUR  100  V(4,0)
.PLOT  TRAN V(4,0)  V(2,0)
.OPTIONS NOPAGE
.END

****     DIODE MODEL PARAMETERS



         IDEAL

IS       1.00D-14

N          .001


****     INITIAL TRANSIENT SOLUTION      TEMPERATURE =   27.000 DEG C

NODE   VOLTAGE      NODE   VOLTAGE


( 2)    .0000    ( 4)     .0000
```

Figure 12.78 Output File FOURIER.OUT.

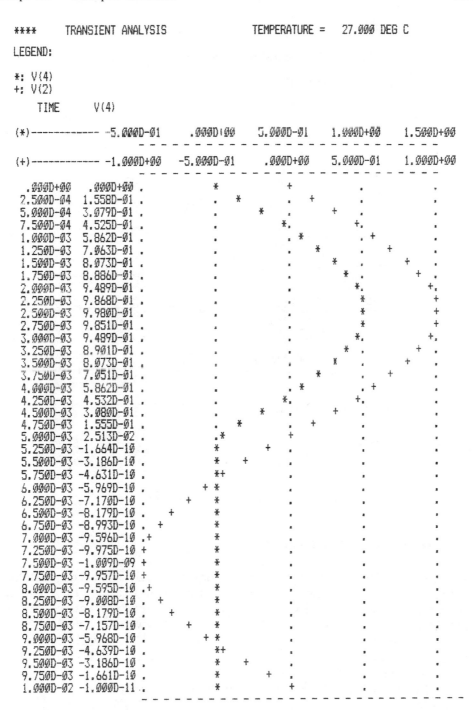

Figure 12.78 (Continued).

```
****      FOURIER ANALYSIS                    TEMPERATURE =    27.000 DEG C

FOURIER COMPONENTS OF TRANSIENT RESPONSE V(4)

DC COMPONENT =    3.178D-01

HARMONIC    FREQUENCY     FOURIER     NORMALIZED     PHASE      NORMALIZED
   NO         (HZ)       COMPONENT    COMPONENT      (DEG)     PHASE (DEG)


    1      1.000D+02    4.990D-01    1.000000     -.067        .000

    2      2.000D+02    2.116D-01     .424020    -90.019      -89.952

    3      3.000D+02    5.181D-04     .001038   -100.573     -100.506

    4      4.000D+02    4.204D-02     .084255    -90.049      -89.982

    5      5.000D+02    5.253D-04     .001053    -93.439      -93.372

    6      6.000D+02    1.782D-02     .035719    -90.157      -90.090

    7      7.000D+02    5.258D-04     .001054    -89.196      -89.129

    8      8.000D+02    9.754D-03     .019549    -90.378      -90.311

    9      9.000D+02    5.249D-04     .001052    -85.983      -85.916


   TOTAL HARMONIC DISTORTION =      43.422851   PERCENT


      JOB CONCLUDED

      TOTAL JOB TIME            13.60
```

Figure 12.78 (Continued).

EXAMPLE 12.6.4, LH0005.CIR

A National Semiconductor LH0005 op-amp will be used to demonstrate the use of the .OP (operating point) and .TF (transfer function) control lines. The equivalent circuit of the LH0005 is shown in Fig. 12.79. Not shown in the schematic are two input voltage sources, VIN1 and VIN2, which are connected to nodes 1 and 3, respectively. Both VIN1 and VIN2 are dead sources, each with 0 V amplitude.

The input file, Fig. 12.80, contains the control line

.TF V(12) VIN1

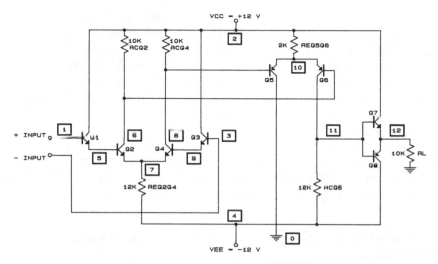

Figure 12.79 Equivalent Circuit of the LH0005 Op-Amp. (Courtesy of National Semiconductor Corp.)

```
LH0005.CIR    NATIONAL SEMICONDUCTOR OP-AMP IC
VCC       2   0   12
VEE       4   0   -12
Q1        2   1   5    MOD1
Q2        6   5   7    MOD1
Q3        2   3   9    MOD1
Q4        8   9   7    MOD1
RCQ2      2   6   10K
RCQ4      2   8   10K
VIN1      1   0   0
VIN2      3   0   0
REQ2Q4    7   4   12K
Q5        0   8   10   MOD2
Q6        11  6   10   MOD2
REQ5Q6    2   10  2K
RCQ6      11  4   12K
Q7        2   11  12   MOD1
Q8        4   11  12   MOD2
RL        12  0   10K
.MODEL    MOD1   NPN
.MODEL    MOD2   PNP
.OP
.TF   V(12)   VIN1
.OPTIONS NOPAGE
.END
```

Figure 12.80   Input File LH0005.CIR.

which will give the DC small-signal input resistance at VIN1,
the output resistance at node 12, and the open-loop DC voltage gain from VIN1 to the op-amp output at node 12. The .OP
control line causes SPICE to print detailed operating point
information about semiconductor junction voltages, currents,
and other parameters. Since the transfer function input is
VIN1, the DC voltage gain that results will be the open-loop
differential gain.

Figure 12.81 contains the output file which provides the
operating point information and transfer function results.
The open-loop differential gain is 3,753 V/V, which agrees
well with data sheet information. If you wanted to determine

```
********11/22/87******** Demo PSpice (May 1986) *******19:46:27********


LH0005.CIR   NATIONAL SEMICONDUCTOR OP-AMP IC


****     CIRCUIT DESCRIPTION


*************************************************************************


       VCC    2   0   12
       VEE    4   0   -12
       Q1     2   1   5   MOD1
       Q2     6   5   7   MOD1
       Q3     2   3   9   MOD1
       Q4     8   9   7   MOD1
       RCQ2   2   6   10K
       RCQ4   2   8   10K
       VIN1   1   0   0
       VIN2   3   0   0
       REQ2Q4 7   4   12K
       Q5     0   8   10  MOD2
       Q6     11  6   10  MOD2
       REQ5Q6 2   10  2K
       RCQ6   11  4   12K
       Q7     2   11  12  MOD1
       Q8     4   11  12  MOD2
       RL     12  0   10K
       .MODEL  MOD1  NPN
       .MODEL  MOD2  PNP
       .OP
       .TF   V(12)   VIN1
       .OPTIONS NOPAGE
       .END
```

Figure 12.81 Output File LH0005.OUT.

```
****      BJT MODEL PARAMETERS


          MOD1      MOD2

TYPE      NPN       PNP

IS        1.00D-16  1.00D-16

BF        100.000   100.000

NF        1.000     1.000

BR        1.000     1.000

NR        1.000     1.000


****      SMALL SIGNAL BIAS SOLUTION       TEMPERATURE =   27.000 DEG C

   NODE   VOLTAGE      NODE   VOLTAGE     NODE   VOLTAGE     NODE   VOLTAGE

(  1)      .0000    (  2)   12.0000    (  3)      .0000   (  4)  -12.0000

(  5)     -.6335    (  6)    7.7086    (  7)   -1.3863    (  0)    7.7086

(  9)     -.6335    ( 10)    8.4792    ( 11)   -1.5325    ( 12)    -.8231


       VOLTAGE SOURCE CURRENTS

       NAME       CURRENT

       VCC       -2.627D-03

       VEE        1.838D-03

       VIN1      -4.334D-08

       VIN2      -4.334D-08


       TOTAL POWER DISSIPATION   5.36D-02  WATTS

****      OPERATING POINT INFORMATION     TEMPERATURE =   27.000 DEG C
```

Figure 12.81 (Continued).

**** BIPOLAR JUNCTION TRANSISTORS

	Q1	Q2	Q3	Q4	Q5	Q6	Q7
MODEL	MOD1	MOD1	MOD1	MOD1	MOD2	MOD2	MOD1
IB	4.33E-08	4.38E-06	4.33E-08	4.38E-06	-8.71E-06	-8.71E-06	-1.35E-11
IC	4.34E-06	4.38E-04	4.34E-06	4.38E-04	-8.71E-04	-8.71E-04	2.64E-11
VBE	.633	.753	.633	.753	-.771	-.771	-.709
VBC	-12.0	-8.34	-12.0	-8.34	7.71	9.24	-13.5
VCE	12.6	9.09	12.6	9.09	-8.48	-10.0	12.8
BETADC	100.	100.	100.	100.	100.	100.	-1.95
GM	1.68E-04	1.69E-02	1.68E-04	1.69E-02	3.37E-02	3.37E-02	.00E+00
RPI	5.96E+05	5.91E+03	5.96E+05	5.91E+03	2.97E+03	2.97E+03	1.00E+14
RX	.00E+00	.00E+00	.00E+00	.00E+00	.00E+00	.00E+00	.00E+00
RO	1.00E+12	1.00E+12	1.00E+12	1.00E+12	1.00E+12	1.00E+12	1.00E+12
CPI	.00E+00	.00E+00	.00E+00	.00E+00	.00E+00	.00E+00	.00E+00
CMU	.00E+00	.00E+00	.00E+00	.00E+00	.00E+00	.00E+00	.00E+00
CBX	.00E+00	.00E+00	.00E+00	.00E+00	.00E+00	.00E+00	.00E+00
CCS	.00E+00	.00E+00	.00E+00	.00E+00	.00E+00	.00E+00	.00E+00
BETAAC	100.	100.	100.	100.	100.	100.	.000
FT	2.67E+15	2.69E+17	2.67E+15	2.69E+17	5.36E+17	5.36E+17	.00E+00

	Q8
MODEL	MOD2
IB	-8.15E-07
IC	-8.15E-05
VBE	-.709
VBC	10.5
VCE	-11.2
BETADC	100.
GM	3.15E-03
RPI	3.17E+04
RX	.00E+00
RO	1.00E+12
CPI	.00E+00
CMU	.00E+00
CBX	.00E+00
CCS	.00E+00
BETAAC	100.
FT	5.02E+16

****   SMALL-SIGNAL CHARACTERISTICS

V(12)/VIN1   = 3.753D+03

INPUT RESISTANCE AT VIN1   = 2.375D+06

OUTPUT RESISTANCE AT V(12)   = 4.150D+02

JOB CONCLUDED

TOTAL JOB TIME   10.70

Figure 12.81 (Continued).

the common-mode gain, the input file shown in Fig. 12.82 would do that.   The input voltage source VIN-CM is connected to node 1 and a 1E-6 ohm resistor, R-SHORT, connects node 1 with node 3.   This makes the input voltage be a common-mode input, so the transfer function gain that results is a common-mode gain.   Its value is 1.132 V/V, which results in a common-mode rejection ratio of 3753/1.132, or 70.4 dB,

EXAMPLE 12.6.5, DISTO.CIR

A BJT differential amplifier, Fig. 12.83, will have a distortion analysis done in conjunction with an AC analysis.   Since the distortion analysis causes a substantial amount of information to be written to the output file for each frequency in the AC analysis, the AC analysis in this example is done at 1 KHz only.   Refer to Fig. 12.84, the input file DISTO.CIR.   In the control line   .DISTO RC1 1, RC1 is the output resistor into which all the distortion power products are to be calculated and the 1 is the interval which tells SPICE to print out the contributions of all nonlinear devices to the distortion

```
LH0005CM.CIR    NATIONAL SEMICONDUCTOR OP-AMP IC
*        COMMON-MODE INPUT, VIN-CM
VCC      2  0  12
VEE      4  0  -12
Q1       2  1  5  MOD1
Q2       6  5  7  MOD1
Q3       2  3  9  MOD1
Q4       8  9  7  MOD1
RCQ2     2  6  10K
RCQ4     2  8  10K
VIN-CM   1  0  0
R-SHORT  1  3  1U
*        R-SHORT IS A JUMPER (1E-6 OHM) TO TIE INPUTS TOGETHER
REQ2Q4   7  4  12K
Q5       0  8  10  MOD2
Q6       11  6  10  MOD2
REQ5Q6   2  10  2K
RCQ6     11  4  12K
Q7       2  11  12  MOD1
Q8       4  11  12  MOD2
RL       12  0  10K
.MODEL  MOD1  NPN
.MODEL  MOD2  PNP
.OP
.TF  V(12)  VIN-CM
.OPTIONS NOPAGE
.END
```

Figure 12.82   Input File LH0005CM.CIR.

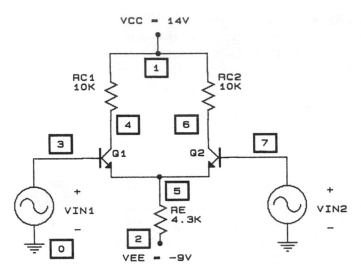

Figure 12.83 Differential Amplifier Circuit.

at each frequency in the AC analysis.   If interval were set to 5, then only at every fifth frequency would that distortion information be printed in the output file.

In the .PRINT control line, HD2 is the second harmonic distortion (2 KHz), HD3 is the third harmonic distortion (3 KHz), SIM2 is the distortion at f1 + f2 (1.9 KHz), DIM2 is the distortion at f1 - f2 (0.1 KHz) and DIM3 is the distortion at 2f1 - f2 (1.1 KHz).

The output file, Fig. 12.85, is printed 133 columns wide even though the default width is 80 columns.   This anomaly

```
DISTO.CIR    DIFF-AMP WITH DISTORTION ANALYSIS
VCC  1  0  14
VEE  2  0  -9
VIN1 3  0  AC  1M
Q1   4  3  5  MOD1
RC1  1  4  10K
RE   5  2  4.3K
VIN2 7  0  0
Q2   6  7  5  MOD1
RC2  1  6  10K
.MODEL  MOD1  NPN(BF=75)
.DISTO  RC1  1
.PRINT AC VM(4)
.PRINT DISTO HD2  HD3  SIM2
.PRINT DISTO DIM2  DIM3
.AC LIN  1  1K  1K
.OPTIONS NOPAGE
.END
```

Figure 12.84 Input File DISTO.CIR.

```
********11-14-87******* ALLSPICE (2.11) 09/18/86 *******17:13:10*******

DISTO.CIR   DIFF-AMP WITH DISTORTION ANALYSIS

****     INPUT LISTING                   TEMPERATURE =   27.000 DEG C

***********************************************************************


  VCC  1  0  14
  VEE  2  0  -9
  VIN1 3  0  AC  1M
  Q1   4  3  5  MOD1
  RC1  1  4  10K
  RE   5  2  4.3K
  VIN2 7  0  0
  Q2   6  7  5  MOD1
  RC2  1  6  10K
  .MODEL MOD1  NPN(BF=75)
  .DISTO RC1  1
  .PRINT AC VM(4)
  .PRINT DISTO  HD2  HD3  SIM2
  .PRINT DISTO  DIM2  DIM3
  .AC LIN  1  1K  1K
  .OPTIONS NOPAGE
  .END
****      BJT MODEL PARAMETERS            TEMPERATURE =   27.000 DEG C



             MOD1
  TYPE       NPN
  IS      1.000D-16
  BF        75.000
  NF         1.000
  BR         1.000
  NR         1.000
****     SMALL SIGNAL BIAS SOLUTION       TEMPERATURE =   27.000 DEG C

  NODE   VOLTAGE    NODE   VOLTAGE    NODE   VOLTAGE    NODE   VOLTAGE


  ( 1)   14.0000   ( 2)   -9.0000   ( 3)    .0000    ( 4)   4.5593

  ( 5)    -.7727   ( 6)   4.5593    ( 7)    .0000



      VOLTAGE SOURCE CURRENTS

        NAME       CURRENT


        VCC      -1.888D-03

        VEE       1.913D-03

        VIN1     -1.259D-05

        VIN2     -1.259D-05


      TOTAL POWER DISSIPATION   4.37D-02  WATTS
****      OPERATING POINT INFORMATION     TEMPERATURE =   27.000 DEG C
```

Figure 12.85 Output File DISTO.OUT.

**** BIPOLAR JUNCTION TRANSISTORS

```
              Q1        Q2
MODEL       MOD1      MOD1
IB        1.26E-05  1.26E-05
IC        9.44E-04  9.44E-04
VBE          .773      .773
VBC       -4.559    -4.559
VCE        5.332     5.332
BETADC    75.000    75.000
GM        3.65E-02  3.65E-02
RPI       2.05E+03  2.05E+03
RX         .00E+00   .00E+00
RO        1.00E+12  1.00E+12
CPI        .00E+00   .00E+00
CMU        .00E+00   .00E+00
CBX        .00E+00   .00E+00
CCS        .00E+00   .00E+00
BETAAC    75.000    75.000
FT        5.81E+17  5.81E+17
****    DISTORTION ANALYSIS          TEMPERATURE =   27.000 DEG C
```

2ND HARMONIC DISTORTION   FREQ1 = 1.00D+03 HZ

DISTORTION FREQUENCY   2.00D+03 HZ   MAG 1.831D-01 PHS   .00

BJT DISTORTION COMPONENTS

NAME		GM	GPI	GO	GMU	GMO2	CB	CBR	CJE	CJC	TOTAL
Q1	MAG	6.018D-02	7.767D-04	2.209D-07	1.000D-20	1.207D-09	1.000D-20	1.000D-20	1.000D-20	1.000D-20	5.940D-02
	PHS	.00	180.00	180.00	.00	180.00	.00	.00	.00	.00	.00
Q2	MAG	5.752D-02	7.670D-04	2.088D-07	1.000D-20	1.147D-09	1.000D-20	1.000D-20	1.000D-20	1.000D-20	5.829D-02
	PHS	180.00	180.00	.00	.00	.00	.00	.00	.00	.00	180.00

HD2   MAGNITUDE 1.106D-03   PHASE   .00   = -59.12 DB
\   3RD HARMONIC DISTORTION   FREQ1 = 1.00D+03 HZ

DISTORTION FREQUENCY   3.00D+03 HZ   MAG 1.831D-01 PHS   .00

BJT DISTORTION COMPONENTS

NAME		GM	GPI	GO	GMU	GMO2	CB	CBR	CJE	CJC	GM2O3	GMO23	TOTAL
Q1	MAG	9.368D-03	1.209D-04	6.382D-06	1.000D-20	1.399D-10	1.000D-20	1.000D-20	1.000D-20	1.000D-20	1.429D-10	5.231D-08	9.254D-03
	PHS	180.00	.00	180.00	.00	.00	.00	.00	.00	.00	180.00	180.00	180.00
Q2	MAG	9.069D-03	1.209D-04	5.964D-06	1.000D-20	1.355D-10	1.000D-20	1.000D-20	1.000D-20	1.000D-20	1.350D-10	4.915D-08	9.196D-03
	PHS	180.00	180.00	180.00	.00	.00	.00	.00	.00	.00	180.00	180.00	180.00

HD3   MAGNITUDE 1.845D-02   PHASE 180.00 = -34.68 DB
\   2ND ORDER INTERMODULATION DIFFERENCE COMPONENT   FREQ1 = 1.00D+03 HZ   FREQ2 = 9.00D+02 HZ

DISTORTION FREQUENCY   1.00D+02 HZ   MAG 1.831D-01 PHS   .00   MAG 1.831D-01 PHS   .00

Figure 12.85 (Continued).

```
BJT DISTORTION COMPONENTS

NAME        GM        GPI        GO        GMU       GMO2      CB        CBR       CJE       CJC       TOTAL
Q1    MAG 1.204D-01 1.553D-03 4.417D-07 1.000D-20 2.414D-09 1.000D-20 1.000D-20 1.000D-20 1.000D-20 1.188D-01
      PHS   .00     180.00    180.00      .00     180.00      .00       .00       .00       .00       .00
Q2    MAG 1.150D-01 1.534D-03 4.177D-07 1.000D-20 2.295D-09 1.000D-20 1.000D-20 1.000D-20 1.000D-20 1.166D-01
      PHS 180.00    180.00      .00       .00       .00       .00       .00       .00       .00     180.00


    IM2D   MAGNITUDE  2.213D-03   PHASE    .00    =  -53.10  DB
\    2ND ORDER INTERMODULATION SUM COMPONENT           FREQ1 =  1.00D+03  HZ           FREQ2 = 9.00D+02  HZ

    DISTORTION FREQUENCY  1.90D+03  HZ          MAG 1.831D-01  PHS    .00      MAG 1.831D-01  PHS    .00




BJT DISTORTION COMPONENTS

NAME        GM        GPI        GO        GMU       GMO2      CB        CBR       CJE       CJC       TOTAL
Q1    MAG 1.204D-01 1.553D-03 4.417D-07 1.000D-20 2.414D-09 1.000D-20 1.000D-20 1.000D-20 1.000D-20 1.188D-01
      PHS   .00     180.00    180.00      .00     180.00      .00       .00       .00       .00       .00
Q2    MAG 1.150D-01 1.534D-03 4.177D-07 1.000D-20 2.295D-09 1.000D-20 1.000D-20 1.000D-20 1.000D-20 1.166D-01
      PHS 180.00    180.00      .00       .00       .00       .00       .00       .00       .00     180.00


    IM2S   MAGNITUDE  2.213D-03   PHASE    .00    =  -53.10  DB
\    3RD ORDER INTERMODULATION DIFFERENCE COMPONENT     FREQ1 =  1.00D+03 · HZ          FREQ2 = 9.00D+02  HZ

    DISTORTION FREQUENCY  1.10D+03  HZ          MAG 1.831D-01  PHS    .00      MAG 1.831D-01  PHS    .00




BJT DISTORTION COMPONENTS

NAME        GM        GPI        GO        GMU       GMO2      CB        CBR       CJE       CJC       GM2O3      GMO23     TOTAL
Q1    MAG 2.811D-02 3.628D-04 1.915D-05 1.000D-20 4.17AD-10 1.000D-20 1.000D-20 1.000D-20 1.000D-20 4.287D-10 1.569D-07 2.776D-02
      PHS 180.00      .00     180.00      .00       .00       .00       .00       .00       .00     180.00    180.00    180.00
Q2    MAG 2.721D-02 3.628D-04 1.789D-05 1.000D-20 4.065D-10 1.000D-20 1.000D-20 1.000D-20 1.000D-20 4.051D-10 1.475D-07 2.759D-02
      PHS 180.00    180.00    180.00      .00       .00       .00       .00       .00       .00     180.00    180.00    180.00


    IM3    MAGNITUDE  5.535D-02   PHASE  180.00    =  -25.14  DB




    APPROXIMATE CROSS MODULATION COMPONENTS

    CMA    MAGNITUDE  2.214D-01                   =  -13.10  DB

    CMP    MAGNITUDE  2.712D-17                   = -331.33  DB
****       AC ANALYSIS                TEMPERATURE =   27.000 DEG C


    FREQ      VM(4)
  1.000E+03   1.831E-01
****       DISTORTION ANALYSIS         TEMPERATURE =   27.000 DEG C


    FREQ      HD2        HD3        SIM2
  1.000E+03   1.106E-03  1.845E-02  2.213E-03
****       DISTORTION ANALYSIS         TEMPERATURE =   27.000 DEG C


    FREQ      DIM2       DIM3
  1.000E+03   2.213E-03  5.535E-02

        JOB CONCLUDED
        TOTAL CPU TIME           7.20
```

Figure 12.85 (Continued).

evidently is necessary so that the large amount of information from the distortion analysis will be in a readable format. The default reference power level is 1 mW, or 0 dBm, and all distortion products are calculated based on that input power level. If the reference power level were made higher, all the distortion products would be larger.

EXAMPLE 12.6.6, SENS.CIR

SPICE has the capability to determine the small-signal DC sensitivity of an output variable (node voltage or current through a voltage source) to all parameters in the circuit. This can be useful for determining which parts in a circuit must be very close to their design values and such parameters as power supply rejection ratio in an amplifier circuit. Figure 12.86 shows the same differential amplifier that was used in Example 12.6.5 for a distortion analysis.

Input file SENS.CIR in Fig. 12.87 includes the control line .SENS V(4), which instructs SPICE to print out the name and the sensitivities (measured both in V/unit and V/percent) of each element in the circuit which will affect the DC volt-

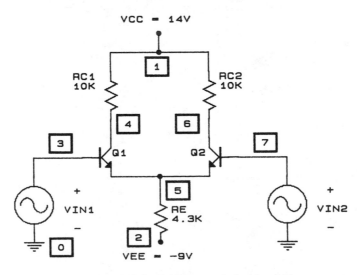

Figure 12.86 Differential Amplfier Circuit.

```
SENS.CIR    DIFF-AMP WITH SENSITIVITY ANALYSIS
VCC  1  0  14
VEE  2  0  -9
VIN1 3  0  AC  1M
Q1   4  3  5  MOD1
RC1  1  4  10K
RE   5  2  4.3K
VIN2 7  0  0
Q2   6  7  5  MOD1
RC2  1  6  10K
.OP
.SENS V(4)
.MODEL MOD1 NPN(BF=75)
.OPTIONS NOPAGE
.END
```

Figure 12.87   Input File SENS.CIR.

age at node 4. If the output variable in the .SENS line was a current, the sensitivity would be expressed in A/unit and A/percent.

An examination of the output file, Fig. 12.88, reveals the results of the sensitivity analysis. As an example of what the results mean, let's look at the effect of RE, the emitter resistor on V(4). The sensitivity is 2.189D-03 V/unit, or 2.189 mV/ohm since RE has units of ohms. This means that if RE increases by one ohm, then V(4) will increase by 2.189 mV.

Since components frequently have tolerances expressed in percent, the RE sensitivity of 9.411D-02 V/percent is a useful result as well. If RE increases by 1%, or 43 ohms, then V(4) will increase by 94.11 mV. Note that 94.11 mV/2.189 mV = 43.

The sensitivity to VIN1 and VIN2 is -183.1 and 181.9, respectively. These are the inverting and non-inverting gains of the differential amplifier. The fact that these gain magnitudes are slightly different accounts for the small, but non-zero, common-mode gain of a differential amplifier. The reason that the normalized sensitivity, in V/percent, is zero is that the DC value of VIN1 and VIN2 are both zero.

Significant insight can be obtained from the sensitivity analysis results. The sensitivity of V(4) to RC1 is 200 million times larger than the sensitivity to RC2, since

9.441D-02/4.644D-10 = 203E6

********11/22/87******** Demo PSpice (May 1986) *******19:52:31********

SENS.CIR   DIFF-AMP WITH SENSITIVITY ANALYSIS

****   CIRCUIT DESCRIPTION

****

```
VCC  1  0  14
VEE  2  0  -9
VIN1 3  0  AC  1M
Q1   4  3  5  MOD1
RC1  1  4  10K
RE   5  2  4.3K
VIN2 7  0  0
Q2   6  7  5  MOD1
RC2  1  6  10K
.OP
.SENS V(4)
.MODEL MOD1 NPN(BF=75)
.OPTIONS NOPAGE
.END
```

****   BJT MODEL PARAMETERS

	MOD1
TYPE	NPN
IS	1.00D-16
BF	75.000
NF	1.000
BR	1.000
NR	1.000

Figure 12.88 Output File SENS.OUT.

```
****     SMALL SIGNAL BIAS SOLUTION      TEMPERATURE =   27.000 DEG C

   NODE   VOLTAGE      NODE   VOLTAGE     NODE   VOLTAGE    NODE   VOLTAGE


 (  1)   14.0000    (  2)   -9.0000    (  3)     .0000   (  4)    4.5593

 (  5)    -.7727    (  6)    4.5593    (  7)     .0000
```

```
     VOLTAGE SOURCE CURRENTS

     NAME        CURRENT


     VCC       -1.888D-03

     VEE        1.913D-03

     VIN1      -1.259D-05

     VIN2      -1.259D-05


     TOTAL POWER DISSIPATION   4.37D-02  WATTS
```

```
****     OPERATING POINT INFORMATION      TEMPERATURE =    27.000 DEG C
```

```
**** BIPOLAR JUNCTION TRANSISTORS
```

```
                 Q1        Q2

     MODEL     MOD1      MOD1
     IB        1.26E-05  1.26E-05
     IC        9.44E-04  9.44E-04
     VBE        .773      .773
     VBC       -4.56     -4.56
     VCE        5.33      5.33
     BETADC    75.0      75.0
     GM        3.65E-02  3.65E-02
     RPI       2.05E+03  2.05E+03
     RX         .00E+00   .00E+00
     RO        1.00E+12  1.00E+12
     CPI        .00E+00   .00E+00
     CMU        .00E+00   .00E+00
     CBX        .00E+00   .00E+00
     CCS        .00E+00   .00E+00
     BETAAC    75.0      75.0
     FT        5.81E+17  5.81E+17
```

Figure 12.88 (Continued).

```
****     DC SENSITIVITY ANALYSIS              TEMPERATURE =   27.000 DEG C

DC SENSITIVITIES OF OUTPUT V(4)


          ELEMENT          ELEMENT          ELEMENT         NORMALIZED
          NAME             VALUE            SENSITIVITY     SENSITIVITY
                                            (VOLTS/UNIT)   (VOLTS/PERCENT)


          RC1              1.000D+04        -9.441D-04      -9.441D-02
          RE               4.300D+03         2.189D-03       9.411D-02
          RC2              1.000D+04        -4.644D-12      -4.644D-10
          VCC              1.400D+01         1.000D+00       1.400D-01
          VEE             -9.000D+00         1.144D+00      -1.030D-01
          VIN1             .000D+00         -1.831D+02       .000D+00
          VIN2             .000D+00          1.819D+02       .000D+00
    Q1
          RB               0.               0.              0.
          RC               0.               0.              0.
          RE               0.               0.              0.
          BF               7.500D+01        -8.255D-04      -6.192D-04
          ISE              0.               0.              0.
          BR               1.000D+00         1.000D-12       1.000D-14
          ISC              0.               0.              0.
          IS               1.000D-16        -4.735D+16      -4.735D-02
          NE               1.500D+00         .000D+00        .000D+00
          NC               2.000D+00         .000D+00        .000D+00
          IKF              0.               0.              0.
          IKR              0.               0.              0.
          VAF              0.               0.              0.
          VAR              0.               0.              0.
    Q2
          RB               0.               0.              0.
          RC               0.               0.              0.
          RE               0.               0.              0.
          BF               7.500D+01        -8.255D-04      -6.192D-04
          ISE              0.               0.              0.
          BR               1.000D+00         4.919D-21       4.919D-23
          ISC              0.               0.              0.
          IS               1.000D-16         4.706D+16       4.706D-02
          NE               1.500D+00         .000D+00        .000D+00
          NC               2.000D+00         .000D+00        .000D+00
          IKF              0.               0.              0.
          IKR              0.               0.              0.
          VAF              0.               0.              0.
          VAR              0.               0.              0.



    JOB CONCLUDED

    TOTAL JOB TIME            7.10
```

Figure 12.88 (Continued).

EXAMPLE 12.6.7, NOISE.CIR

In Fig. 12.89, the last differential amplifier to appear in this chapter, a noise analysis will be done. Noise analysis can only be done along with an AC analysis. SPICE must be told across which output nodes the noise voltage is assumed to appear, and which input independent source (voltage or current) is to be used for the input noise reference.

The input file, Fig. 12.90, instructs SPICE to perform an AC analysis from 1 Hz to 100 MHz with one frequency per de-

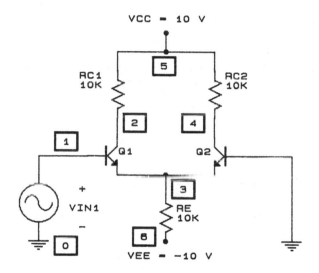

Figure 12.89 Differential Amplifier Circuit.

```
NOISE.CIR   DIFFERENTIAL AMPLIFIER WITH NOISE ANALYSIS
VIN1 1 0 AC 10M
Q1 2 1 3 MOD1
Q2 4 0 3 MOD1
.MODEL MOD1 NPN(RB=2  RC=.8  RE=.5  KF=1E-12)
VCC 5 0 10
VEE 6 0 -10
RC1 5 2 10K
RC2 5 4 10K
RE 3 6 10K
.AC DEC 1 1 100MEG
.NOISE V(2,3) VIN1  8
.OPTIONS NOPAGE
.PRINT AC VM(2)
.PLOT NOISE ONOISE  INOISE
.END
```

Figure 12.90 Input File NOISE.CIR.

cade, for a total of nine frequencies. The .NOISE control line specifies V(2,3) as the assumed output, VIN1 as the input noise reference and that the summary printout of all parameters that contribute to noise (resistors and semiconductors) be printed at every eighth frequency. A graph of input and output noise versus frequency is requested by the .PLOT control line.

Output file NOISE.OUT, Fig. 12.91, contains a summary printout of the noise at 1 Hz and 100 MHz. All output noise contributions are the same at both frequencies except for the flicker noise of the transistors. The input noise multiplied by the voltage gain (transfer function value of 88.41 V/V) gives the output noise.

The last item in the output file is the graph of input noise and output noise versus frequency, which shows that the flicker noise is significant below 100 KHz.

```
********11/21/87********* Demo PSpice (May 1986) *******20:08:17********


NOISE.CIR   DIFFERENTIAL AMPLIFIER WITH NOISE ANALYSIS


****      CIRCUIT DESCRIPTION


****************************************************************************


VIN1 1 0 AC 10M
Q1 2 1 3 MOD1
Q2 4 0 3 MOD1
.MODEL MOD1 NPN(RB=2  RC=.8  RE=.5  KF=1E-12)
VCC 5 0 10
VEE 6 0 -10
RC1 5 2 10K
RC2 5 4 10K
RE 3 6 10K
.AC DEC 1 1 100MEG
.NOISE V(2,3) VIN1  8
.OPTIONS NOPAGE
.PRINT AC VM(2)
.PLOT NOISE ONOISE  INOISE
.END
```

Figure 12.91 Output File NOISE.OUT.

```
****      BJT MODEL PARAMETERS


          MOD1

TYPE      NPN

IS        1.00D-16

BF        100.000

NF          1.000

BR          1.000

NR          1.000

RB          2.000

RE           .500

RC           .800

KF        1.00D-12

****      SMALL SIGNAL BIAS SOLUTION      TEMPERATURE =   27.000 DEG C

   NODE   VOLTAGE       NODE   VOLTAGE      NODE   VOLTAGE       NODE   VOLTAGE

 (  1)     .0000      (  2)    5.4229     (  3)   -.7542      (  4)    5.4229

 (  5)   10.0000      (  6)  -10.0000

****      NOISE ANALYSIS                  TEMPERATURE =   27.000 DEG C


     FREQUENCY =   1.000D+00 HZ



**** RESISTOR SQUARED NOISE VOLTAGES (SQ V/HZ)



          RC1        RC2        RE
TOTAL    1.658D-16 4.208D-33 3.993D-17
```

Figure 12.91 (Continued).

```
**** TRANSISTOR SQUARED NOISE VOLTAGES (SQ V/HZ)


               Q1          Q2
RB        2.591D-16 2.505D-16

RC        3.074D-36 3.367D-37

RE        6.479D-17 6.262D-17

IB        3.859D-17 3.228D-17

IC        3.737D-15 3.471D-15

FN        1.204D-10 1.007D-10

TOTAL     1.204D-10 1.007D-10



**** TOTAL OUTPUT NOISE VOLTAGE         = 2.212D-10 SQ V/HZ

                                        = 1.487D-05 V/RT HZ

     TRANSFER FUNCTION VALUE:

        V(2,3)/VIN1                      = 8.841D+01

     EQUIVALENT INPUT NOISE AT VIN1      = 1.682D-07 V/RT HZ

****      NOISE ANALYSIS                    TEMPERATURE =   27.000 DEG C

     FREQUENCY =  1.000D+08 HZ



**** RESISTOR SQUARED NOISE VOLTAGES (SQ V/HZ)


            RC1        RC2        RE
TOTAL     1.658D-16 4.208D-33 3.993D-17
```

Figure 12.91 (Continued).

**** TRANSISTOR SQUARED NOISE VOLTAGES (SQ V/HZ)

	Q1	Q2
RB	2.591D-16	2.505D-16
RC	3.074D-36	3.367D-37
RE	6.479D-17	6.262D-17
IB	3.859D-17	3.228D-17
IC	3.737D-15	3.471D-15
FN	1.204D-18	1.007D-18
TOTAL	4.100D-15	3.818D-15

**** TOTAL OUTPUT NOISE VOLTAGE   = 8.124D-15 SQ V/HZ

                                         = 9.013D-08 V/RT HZ

    TRANSFER FUNCTION VALUE:

        V(2,3)/VIN1   = 8.841D+01

    EQUIVALENT INPUT NOISE AT VIN1   = 1.019D-09 V/RT HZ

****   AC ANALYSIS   TEMPERATURE =   27.000 DEG C

FREQ	VM(2)
1.000E+00	8.792E-01
1.000E+01	8.792E-01
1.000E+02	8.792E-01
1.000E+03	8.792E-01
1.000E+04	8.792E-01
1.000E+05	8.792E-01
1.000E+06	8.792E-01
1.000E+07	8.792E-01
1.000E+08	8.792E-01

Figure 12.91 (Continued).

```
****     AC ANALYSIS                     TEMPERATURE =   27.000 DEG C


LEGEND:

*: ONOISE
+: INOISE

   FREQ      ONOISE

(*)------------ 1.000D-08    1.000D-07    1.000D-06    1.000D-05    1.000D-04
                - - - - - - - - - - - - - - - - - - - - - - - - - - - -

(+)------------ 1.000D-09    1.000D-08    1.000D-07    1.000D-06    1.000D-05
                - - - - - - - - - - - - - - - - - - - - - - - - - - - -
  1.000D+00  1.487D-05 .          .          . +        . *        .
  1.000D+01  4.704D-06 .          .        + .        * .          .
  1.000D+02  1.490D-06 .          . +      . *          .          .
  1.000D+03  4.788D-07 .       + .        * .          .          .
  1.000D+04  1.739D-07 .   +      . *        .          .          .
  1.000D+05  1.017D-07 .+        *          .          .          .
  1.000D+06  9.134D-08 +        *.          .          .          .
  1.000D+07  9.024D-08 +        *.          .          .          .
  1.000D+08  9.013D-08 +        *.          .          .        - - -
                - - - - - - - - - - - - - - - - - - - - - -


  JOB CONCLUDED

  TOTAL JOB TIME            9.50
```

Figure 12.91 (Continued).

CHAPTER SUMMARY

A good reference to use when generating SPICE input files is a similar input file. 30 example files covering every type of analysis were presented in this chapter. Additional example files may be found in Chaps. 4, 9, 10 and 11, and in Appendix B.

Organize the input files you generate, along with your notes about details of element and control lines. These will be good references for future SPICE work you do.

The example circuits in this chapter, with brief descriptions of the SPICE functions they illustrate, follow:

Input File Name   Circuit Type & SPICE Functions Illustrated

DCMESS.CIR	DC circuit	.OP
VANDI.CIR	DC circuit	.OP
DC-CKT.CIR	DC circuit with VCCS	
DI-VA.CIR	Diode V-A characteristic	.DC
FETINV.CIR	JFET inverter	.DC
TTLINV.CIR	7404 TTL logic gate	.DC
LOPASS.CIR	R-C filter	.AC
WEIN.CIR	Wein bridge	.AC
LCMATCH.CIR	Many L, C elements	.AC
SERIESAC.CIR	2 series resonant ckts.	.AC
DUBTUNE.CIR	Double-tuned transformer	.AC
2AC.CIR	2-source problem	.AC
2HP.CIR	2 active filter in parallel	.AC
ACDIFAMP.CIR	BJT differential amp	.AC
TRIANG.CIR	PWL triangle, R-C ckt.	.TRAN
LC.CIR	PWL current pulse into L-C	.TRAN
EKG-LP.CIR	PWL electro-cardiogram	.TRAN
PWRSUPLI.CIR	1/2 wave rectifier,ideal di	.TRAN
CMOSNAND.CIR	MOSFET nand gate	.TRAN
TTL.CIR	BJT 7404 TTL inv. gate	.TRAN
ASTABLE.CIR	2-BJT clock, w/initial cond.	.TRAN,.IC, UIC
ABSVAL.CIR	Prec. rectifier, subcircuits	.TRAN
AM.CIR	Non-linear VCVS, ampl. mod.	.TRAN
VARYTEMP.CIR	Diode V-A graph, 2 temps.	.TEMP, .DC
DELTAT.CIR	BJT diff-amp, transfer func.	.TEMP, .TF
FOURIER.CIR	1/2 wave rect., Fourier anal.	.TRAN, .FOUR
LH0005.CIR	Op-amp, open-loop	.TF, .OP
DISTO.CIR	BJT diff-amp, distortion	.AC, .DISTO
SENS.CIR	BJT diff-amp, sensitivity	.SENS
NOISE.CIR	BJT diff-amp, noise analysis	.AC, .NOISE

ANSWERS TO PROBLEMS

Chapter 4

Problem 1.

```
CHAP4PR1.CIR    3 R'S, 2 BATTERIES
* SOLUTION TO CHAPTER 4, PROBLEM 1.
VL 0 5 10
RL 5 12 2K
RM 12 0 3K
RR 12 38 1K
VR 38 0 20
.OPTIONS NOPAGE
.END
```

Problem 2.

```
CHAP4PR2.CIR  PARALLEL RESONANT CIRCUIT, 10 KHZ
* SOLUTION TO CHAPTER 4, PROBLEM 2.
```

```
R1   1   2   1K
R2   3   0   1
VGEN   1   0   AC   1
L1   2   3   1M
C1   2   0   0.2533U
.AC LIN 21   5000 15000
.PRINT AC VM(2) VP(2)
.PLOT AC  VM(2) VP(2)
.END
```

Problem 3.

```
CHAP4PR3.CIR   R-L CIRCUIT WITH PULSE INPUT
*  SOLUTION TO CHAPTER 4, PROBLEM 3
VIN 1 0 PULSE(0 1 10U 1N 1N 80U)
RA  1  2  1000
LA  2  0  20E-3
.TRAN 5U 160U
.PRINT TRAN V(2) V(1)
.PLOT   TRAN V(2) V(1)
.END
```

Problem 4.

```
CHAP4PR4.CIR     TWO-SOURCE DC BRIDGE
*  SOLUTION TO CHAPTER 4, PROBLEM 4.
V1   14   0   5
R1   14   16   10
R2   16   0   40
R3   14   18   20
R4   18   12   50
R5   16   18   30
V2   12   0   7
.OPTIONS NOPAGE
.END
```

Chapter 6

Problem 1.

   a.

```
BRIDGE.CIR   DC BRIDGE CIRCUIT
*  SOLUTION TO CHAPTER 6, PROBLEM 1.
V    10  0  20
RA  10  12  6K
RB  12  14  2K
RC  12  16  5K
RD  14  16  4K
RE  14   0  3K
RF  16   0  1K
.OP
.OPTIONS NOPAGE
.END
```

   c. V(14) - V(16) = 3.0343 V - 1.3377 V = 1.6966 V.

Problem 2.

   a,c.

```
OH-MY.CIR    DC CIRCUIT WITH A LOT HAPPENING
*  SOLUTION TO CHAPTER 6, PROBLEM 2.
VL   1  0  10
R1   1  2  1
I    0  2  1
R2   2  3  2
VSENSE  4  3  0
R4   2  5  5
E    4  0  2  1  7
R3   4  5  3
VR   5  0  3
.SENS I(VSENSE)
.OP
.OPTIONS NOPAGE
.END
```

   b. I(VSENSE) = 4 A, from node 4 to node 3.

Problem 3.

    a.

```
PRE-EMPH.CIR   A PRE-EMPHASIS CIRCUIT FOR FM
* SOLUTION TO CHAPTER 6, PROBLEM 3.
VIN1  0  AC  1
RA  1  2  47K
CA  1  2  1.6E-9
RB  2  0  5K
.AC DEC 10 200 200K
.PLOT AC VM(2) (0.1,1)
.OPTIONS NOPAGE
.END
```

Problem 4.

    a.

```
NOTCH.CIR    ACTIVE NOTCH FILTER, AC ANALYSIS
* SOLUTION TO CHAPTER 6, PROBLEM 4A.
VIN1  0  AC  1
*     ACTIVE HIGH-PASS FILTER FOLLOWS
C1   1  2  1E-7
C2   2  3  1E-7
R1   3  0  2K
R2   2  5  2K
R3   5  4  818
R4   4  0  2K
*     FIRST OP-AMP MODEL FOLLOWS
RIN1   3  4  1MEG
E1   5  0  3  4  1E5
*     ACTIVE LOW-PASS FILTER FOLLOWS
R5   5  6  2K
R6   6  7  2K
R7   9  8  818
R8   8  0  2K
C3   9  6  0.1U
C4   7  0  0.1U
*     SECOND OP-AMP MODEL FOLLOWS
RIN2   7  8  1MEG
E2   9  0  7  8  1E5
```

```
*    TRUE DIFFERENTIAL AMPLIFIER FOLLOWS
R9  1  10  2K
R10 10  0  2K
R11 9  11  2K
R12 11  12  2K
*    THIRD OP-AMP MODEL FOLLOWS
RIN3   10  11  1MEG
E3  12  0  10  11  1E5
.AC  DEC  15  100  10K
.PLOT  AC  VDB(12)
.OPTIONS  NOPAGE
.END
```

Problem 4.

      b. Notch depth, or gain at 800 Hz., = -13.31 dB.

```
NOTCH2.CIR   ACTIVE NOTCH FILTER, AC ANALYSIS, NARROW BAND
*  SOLUTION TO CHAPTER 6, PROBLEM 4B.
VIN 1  0  AC  1
*    ACTIVE HIGH-PASS FILTER FOLLOWS
C1   1  2  1E-7
C2   2  3  1E-7
R1   3  0  2K
R2   2  5  2K
R3   5  4  818
R4   4  0  2K
*    FIRST OP-AMP MODEL FOLLOWS
RIN1   3  4  1MEG
E1   5  0  3  4  1E5
*    ACTIVE LOW-PASS FILTER FOLLOWS R55  6  2K
R6   6  7  2K
R7   9  8  818
R8   8  0  2K
C3   9  6  0.1U
C4   7  0  0.1U
*    SECOND OP-AMP MODEL FOLLOWS
RIN2   7  8  1MEG
E2   9  0  7  8  1E5
*    TRUE DIFFERENTIAL AMPLIFIER FOLLOWS
R9   1  10  2K
R10 10  0  2K
R11 9  11  2K
R12 11  12  2K
```

```
*    THIRD OP-AMP MODEL FOLLOWS
RIN3   10  11  1MEG
E3   12  0  10  11  1E5
.AC  LIN  31  500  1100
.PLOT  AC  VDB(12)
.OPTIONS  NOPAGE
.END
```

Problem 5.

    a.

```
MANY-C.CIR   7 CAPS, TO DETERMINE EFFECTIVE CAPACITANCE
*  SOLUTION TO CHAPTER 6, PROBLEM 5A.
C1   1  3  2U
C2   1  2  3U
C3   2  3  1U
C4   2  4  1.5U
C5   3  4  2.5U
C6   3  0  3.5U
C7   4  0  0.5U
L    10  1  1MH
V    10  0  AC
*    DUMMY RESISTORS ARE NEEDED TO GIVE DC PATHS TO GROUND
R3   3  0  1G
R4   4  0  1G
R2   2  0  1G
.AC DEC 5 1 1MEG
.PLOT AC VM(1) VP(1)
.OPTIONS NOPAGE
.END
```

    b.

```
MANY-C3.CIR   7 CAPS, TO DETERMINE EFFECTIVE CAPACITANCE
*  SOLUTION TO CHAPTER 6, PROBLEM 5B.
*    NARROWER FREQUENCY RANGE
C1   1  3  2U
C2   1  2  3U
C3   2  3  1U
C4   2  4  1.5U
C5   3  4  2.5U
C6   3  0  3.5U
C7   4  0  0.5U
```

```
L    10  1  1MH
V    10  0  AC
*    DUMMY RESISTORS ARE NEEDED TO GIVE DC PATHS TO GROUND
R3   3  0  1G
R4   4  0  1G
R2   2  0  1G
.AC  LIN  26  3.7K  3.8K
.PLOT AC VM(1) VP(1)
.OPTIONS NOPAGE
.END
```

Problem 6.

     a.

```
HP-RING.CIR   HIGH-PASS ACTIVE FILTER, UNDER-DAMPED, RINGS
*  SOLUTION TO CHAPTER 6, PROBLEM 6.
VIN 1  0  PULSE(0  1  0.5M  1U  1U  0.4M)
C1   1  2  8N
C2   2  3  8N
R1   5  2  2K
R2   3  0  2K
R3   4  0  2K
R4   5  4  3.96K
*    OP-AMP MODEL FOLLOWS
RIN 3  4  1E6
E    5  0  3  4  5E4
.TRAN  0.1M  5M
.PLOT TRAN V(5) V(1)
.OPTIONS NOPAGE
.END
```

Problem 7.

     a.

```
SQ-WAVE.CIR   FOURIER ANALYSIS OF A 1 KHZ SQUAREWAVE
*  SOLUTION TO CHAPTER 6, PROBLEM 7.
V    1  0  PULSE(-1  1  0  1N  1N  0.5M  1M)
R    1  0  1
.TRAN  0.01M  2M
.FOUR  1K  V(1)
.OPTIONS NOPAGE
.END
```

Appendix A

OTHER CONTROL LINES

A.1 INTRODUCTION

Four control lines were not covered previously in this book, since they are the kinds of control lines you can generally do without until you start doing some serious and advanced SPICE analysis. The control lines are: .WIDTH, .OPTIONS, .NODESET and .IC. Browse through this appendix, noting particularly the many choices that can be specified using the .OPTIONS control line. Try those that you wish to know more about, and compare the output files of the analysis with and without the options specified.

A.2 .WIDTH CONTROL LINE

The general form for the .WIDTH control line is

    .WIDTH IN = COLNUMIN OUT = COLNUMOUT

where COLNUMIN is the last column of the input file that SPICE will read. This specification takes effect on the next input file line after .WIDTH IN = COLNUMIN; therefore it is

best to place it just after the title which is the first line
in the input file.

COLNUMOUT is the width of the output file; allowable values are 80 and 133.   On some mainframe versions of SPICE 2G
the default value of COLNUMOUT is 133.   When the output file
is typed on a terminal CRT, lines wider than 80 columns will
wrap around, making the output file hard to interpret.   Thus
it is a good idea to put a control line .WIDTH   OUT = 80 in
an input file which is run on a mainframe computer.   One exception to this is when you desire the maximum resolution possible on a graph generated by a .PLOT line, and will print
that graph on a line printer 133 columns wide.   An output
file with a width of 133 will have a smoother plot with less
quantization error in the graph.

EXAMPLE:

   .WIDTH   OUT = 80
   The output file will have a width of 80 columns.   Note
that it is not necessary to specify an input width as well.

A.3   .OPTIONS CONTROL LINE

The general form for the .OPTIONS control line is

   .OPTIONS OPT1 OPT2 . . .   (or OPT = OPTVAL . . .)

There are 32 options which may be specified; normally only a
few are used in any particular input file.   Options can be
listed in any order.   Many SPICE users will have little need
to modify the default option values.   For more information on
these options, refer to the publications in Appendix E.1.   In
the listing below, "x" stands for a positive number.   The
options and their effects are:

Option	Effect
ACCT	causes accounting and run time statistics to be printed.
LIST	causes the summary listing of the input data to be printed.
NOMOD	suppresses the printout of the model parameters.

Option	Effect
NOPAGE	suppresses page ejects.
NODE	causes the printing of the node table.
OPTS	causes the option values to be printed.
GMIN=x	resets the value of GMIN, the minimum conductance allowed by the program. The default value is 1.0E-12.
RELTOL=x	resets the relative error tolerance of the program. The default value is 0.001 (0.1 percent).
ABSTOL=x	resets the absolute current error tolerance of the program. The default value is 1 picoamp.
VNTOL=x	resets the absolute voltage error tolerance of the program. The default value is 1 microvolt.
TRTOL=x	resets the transient error tolerance. The default value is 7.0. This parameter is an estimate of the factor by which SPICE overestimates the actual truncation error.
CHGTOL=x	resets the charge tolerance of the program. The default value is 1.0E-14.
PIVTOL=x	resets the absolute minimum value for a matrix entry to be accepted as a pivot. The default value is 1.0E-13.
PIVREL=x	resets the relative ratio between the largest column entry and an acceptable pivot value. The default value is 1.0E-3. In the numerical pivoting algorithm the allowed minimum pivot value is determined by: EPSREL = AMAX1(PIVREL*MAXVAL,PIVTOL) where MAXVAL is the maximum element in the column where a pivot is sought (partial pivoting).
NUMDGT=x	resets the number of significant digits printed for output variable values. X must satisfy the relation $0 < x < 8$. The default value is 4. Note: this option is independent of the error tolerance used by SPICE (i.e., if the values of options RELTOL, ABSTOL, etc., are not changed then one may be printing numerical noise for NUMDGT > 4.
TNOM=x	resets the nominal temperature. The default value is 27 deg C (300 deg K).

Option	Effect
ITL1=x	resets the dc iteration limit. The default is 100.
ITL2=x	resets the dc transfer curve iteration limit. The default is 50.
ITL3=x	resets the lower transient analysis iteration limit. The default value is 4.
ITL4=x	resets the transient analysis timepoint iteration limit. The default is 10.
ITL5=x	resets the transient analysis total iteration limit. The default is 5000. Set ITL5 = 0 to omit this test.
CPTIME=x	the maximum cpu-time in seconds allowed for this job.
LIMTIM=x	resets the amount of cpu time reserved by SPICE for generating plots should a cpu time-limit cause job termination. The default value is 2 (seconds).
LIMPTS=x	resets the total number of points that can be printed or plotted in a dc, ac, or transient analysis. The default value is 201.
LVLCOD=x	if x is 2 (two), then machine code for the matrix solution will be generated. Otherwise, no machine code is generated. The default value is 2. Applies only to CDC computers.
LVLTIM=x	if x is 1 (one), the iteration timestep control is used. If x is 2 (two), the truncation-error timestep is used. The default value is 2. If method=Gear and MAXORD>2 then LVLTIM is set to 2 by SPICE.
METHOD=name	sets the numerical integration method used by SPICE. Possible names are Gear or trapezoidal. The default is trapezoidal.
MAXORD=x	sets the maximum order for the integration method if Gear's variable-order method is used. x must be between 2 and 6. The default value is 2.
DEFL=x	resets the value for MOS channel length; the default is 100.0 micrometer.
DEFW=x	resets the value for MOS channel width; the default is 100.0 micrometer.

Option	Effect
DEFAD=x	resets the value for MOS drain diffusion area; the default is 0.0.
DEFAS=x	resets the value for MOS source diffusion area; the default is 0.0.

EXAMPLE:

.OPTIONS NOPAGE NODE LIMPTS = 501

This control line will shorten the output file by suppressing page ejects, will print a node table, and will increase from 201 (the default value) to 501 the maximum number of points that can be plotted or printed in an AC, DC or transient analysis.

EXAMPLE:

.OPTIONS TNOM = 0 ACCT RELTOL = .01

This control line changes the nominal temperature to 0 degrees C, causes the accounting and run-time statistics to be printed and resets the relative error tolerance of the program from .001 (the default value) to .01 (1 percent).

A.4 .NODESET CONTROL LINE

The general form for the .NODESET control line is

.NODESET V(NODENUM1) = VAL1 V(NODENUM2) = VAL2 . . .

where NODENUM is a positive integer representing a node other than node 0, and VAL is the voltage at that node. SPICE will do the initial attempt to find a DC or transient bias point solution with the nodes specified held to the values shown. It is not necessary to specify the voltages at all nodes. Subsequently this restriction is released and SPICE continues to the correct solution.

The .NODESET line may be of use in bi-stable or astable circuits, where the program user can make an educated guess as to certain node voltages.

EXAMPLE:

.NODESET V(12) = -5 V(8) = 12
The voltage at node 12 is set to -5 V, and the node 8 voltage is set to 12 V for the initial bias point solution.

A.5 .IC CONTROL LINE

The general form for the .IC control line is

.IC V(NODENUM1) = VAL1 V(NODENUM2) = VAL2 . . .

where NODENUM is a positive integer representing a node other than node 0, and VAL is the voltage at that node. This line is used to set transient initial conditions. It has two different interpretations, depending on whether the UIC parameter is specified on the .TRAN control line. Also, one should not confuse this line with the .NODESET line (see A.4 above). The .NODESET line is only to help DC convergence, and does not affect final bias solution (except for multistable circuits). The two interpretations of this line are as follows:

1. When the UIC parameter is specified on the .TRAN line, then the node voltages specified on the .IC line are used to compute the capacitor, diode, BJT, JFET and MOSFET initial conditions. This is equivalent to specifying the IC = . . . parameter on each device element line, but is much more convenient. The IC = . . . parameter can still be specified and will take precedence over the .IC values. Since no DC bias (initial transient) solution is computed before the transient analysis, one should take care to specify all DC source voltages on the .IC line if they are to be used to compute device initial conditions.

2. When the UIC parameter is not specified on the .TRAN line, the DC bias (initial transient) solution will be computed before the transient analysis. In this case, the node voltages specified on the .IC line will be forced to the desired initial values during the bias solution. During transient analysis, the constraint on these node voltages is removed.

EXAMPLE:

.IC   V(12) = -5   V(8) = 12

The voltage at node 12 is set to -5 V, and the node 8 voltage is set to 12 V for the final bias point solution if the UIC parameter is not specified on the .TRAN control line. If the UIC parameter is specified, then the voltages for nodes 12 and 8 are used to compute the capacitor, diode, BJT, JFET and MOSFET initial conditions.

Appendix B

TRANSFORMERS AND SWITCHES

B.1 ADVANCED HANDY HINTS

If you have gotten to this point in the book, your appetite for shortcuts, hints and neat ways of doing things with SPICE no doubt is whetted. In this appendix we will cover how to make an ideal transformer, how to make a lossless transformer using SPICE inductors, making switches which are voltage-controlled or current-controlled, and lastly how to make a switched three-phase power source. (Switches are included in SPICE version 3, but not in version 2.)

No doubt as you use SPICE more and more, you will encounter a circuit component which you would like to include in a SPICE analysis but which is not known to SPICE. Examples of such a device might be a mechanical relay, a varistor or a unijunction transistor. By spending some time looking into how devices are modeled, and learning how to use to their fullest capabilities those devices which SPICE does recognize, you will be able to create excellent approximations for most devices.

B.2 IDEAL TRANSFORMER, NO INDUCTORS

The ideal transformer is a fictitious device which divides voltage by a constant called turns ratio, multiplies current by the same ratio and changes impedance connected to it by the square of the turns ratio. Turns ratio is the number of primary turns divided by the number of secondary turns. Real transformers are magnetic machines with inductance, ohmic loss resistance and sometimes substantial losses in the magnetic core. An ideal transformer can be a useful circuit element to use when the inductance and losses of a real transformer can be neglected.

Figure B.1 shows the schematic symbol for an ideal transformer, and the SPICE schematic for its model. It uses two controlled sources. The first is a voltage-controlled voltage source (VCVS), EPRI-SEC, which makes the secondary voltage equal to the product of the primary voltage and 1/(turns ratio). The second is a current-controlled current source (CCCS), FSEC-PRI, which makes the primary current equal to the secondary current multiplied by 1/(turns ratio).

In order to control FSEC-PRI a way of sensing the current flow in the secondary is needed. That is done by V-ISEC, a dead voltage source acting as an ammeter, which measures the current flowing up through EPRI-SEC.

N is the "gain" of the controlled sources, determined by the number of secondary winding turns divided by the number of primary winding turns

N = # secondary turns/(# primary turns) = 1/(turns ratio)

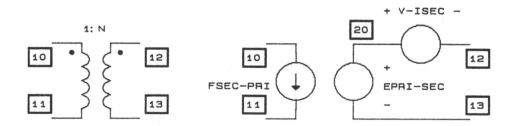

Figure B.1 Ideal Transformer and SPICE Model.

The element lines below describe the ideal transformer model with N having a value of 2. In other words, the transformer is a 2:1 step-up transformer if the primary winding connections are nodes 10 and 11.

```
EPRI-SEC  20   13   10   11   2
FSEC-PRI  10   11   V-ISEC    2
V-ISEC    20   12   0
```

One word of caution: this SPICE circuit for an ideal transformer is a model based on behavior, not physics. It will work with DC sources as well as it will with AC sources. Real transformers do only one thing with DC sources; smoke and burn up.

B.3 IDEAL TRANSFORMER, WITH INDUCTORS

Another way to model a lossless transformer is to use two inductors which have a coefficient of coupling of 1.0. If the inductance of either the primary (LPRI) or the secondary (LSEC) is known, the inductance of the other winding can be determined using the number of primary and secondary turns, as follows

N = # secondary turns/(# primary turns)

$LSEC = LPRI * N^2$, or $LPRI = LSEC/(N^2)$

The transformer and the SPICE model, assuming an N of 5 and primary inductance of 500 mH, is shown in Fig. B.2.
The secondary inductance can be calculated by

$LSEC = 50 \text{ mH} * 5^2 = (5E-2)25 = 1.25 \text{ H}$

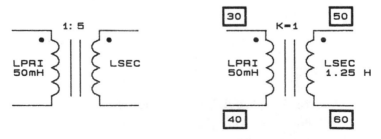

Figure B.2 Transformer and SPICE Model.

The element lines of this transformer model for use in a SPICE input file would be

```
LPRI       30   40   50E-3
LSEC       50   60   1.25
KPRI-SEC  LPRI LSEC      1.00
```

Notice that the first node for each inductor is the node which is connected to the dotted end in the schematic, Fig. B.2. This ensures the proper polarity of induced voltage in the secondary. Unlike the ideal transformer without inductance in B.2, this model with inductance will not work with DC. With DC current flowing through the primary, there is no changing magnetic flux in the core of the transformer, so there is no induced voltage in the secondary.

B.4 MODEL FOR A SWITCH

SPICE version 2 lacks a switch model. A switch turned on at some time after time zero can be used to simulate many things, including the abrupt changing of a load on a circuit such as a power supply or amplifier. Fortunately, making a switch turn on (or off) is relatively simple using a polynomial voltage- or current-controlled current source. See Appendix C for information on polynomial controlled sources.

In this section a VCCS will be explained in detail. The switch can have any value of resistance in the off condition (except 0) and any value of resistance in the on condition (except 0).

The trick which allows the switch to be modeled is that a voltage-controlled current source can be made, through which the current is the mathematical product of two voltages:

The first is a step from 0 V to 1 V, for a switch closing (turning on), or from 1 V to 0V for a switch opening (turning off). This can easily be accomplished with an independent PULSE source or a PWL source.

The second is the voltage across the VCCS itself. By assigning a transconductance of a large value, perhaps 1E6 siemens, the "on" resistance of the switch will be 1E-6 ohms.

In other words, when the first voltage is 0 V, the current is
0 A.   This means infinite resistance in the open condition.
When the first voltage is 1 V, the current through the switch
is the product of 1, the voltage across the switch, and the
very large transconductance value.   The VCCS behaves like a
resistor of a tiny value, which can be specified in the VCCS
element line.

In order to illustrate how to put a switch into a circuit, let's model the switch in Fig. B.3.   The switch will
close at t=1 s, and open at t=2 s.   A little mesh analysis or
nodal analysis or use of the superposition theorem would re-
veal that when the switch is open, the voltage at node 3 is 1
V. With the switch closed, the voltage at node 3 is 2 V.

Figure B.4 shows the switch replaced with a polynomial
VCCS called GSWITCH, along with an independent PULSE voltage
source (VPULSE) which will "operate" the switch.   Since SPICE
is fussy about nodes with only one element connected, a dummy
resistor (one ohm) called RPULSE is connected across VPULSE.

The input file for this example, Fig. B.5, defines the
pulse as starting at 0 V, pulse value 1 V, delayed in time by
1 s, with rise and fall times of 1 ms and pulse width of 1 s.
The VCCS GSWITCH is connected between nodes 1 and 2, and is
controlled by the voltage V(100,0) (the pulse) and the voltage V(1,2) (the nodes to which it is connected).   The p4 coefficient in the GSWITCH element line is specified as 1E6
siemens, a very large transconductance, the inverse of which
is the very small "on" resistance.

In order to test this circuit, a transient analysis is
performed for 2.5 seconds.   The output file, Fig. B.6, shows
a graph of the pulse voltage and the voltage at node 3.   The
voltage at node 3 does indeed jump up from 1 V to 2 V during
the time that the pulse is 1 V (the time that the switch is
"closed").

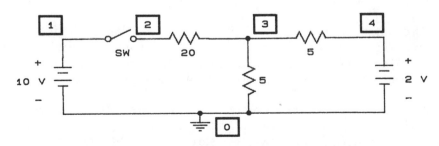

Figure B.3 DC Circuit with Switch.

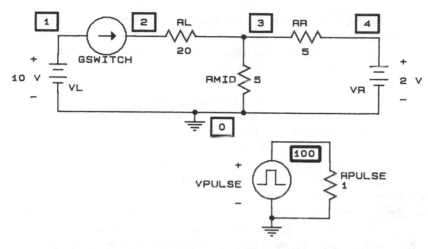

Figure B.4 DC Circuit with VCCS Replacing Switch and a Pulse Voltage Source to Operate Switch.

If the resistance of the switch in the "off" state is to be less than infinity, a single resistor whose value is equal to the off resistance can be placed in parallel with the VCCS acting as the switch.

B.5 SWITCHED THREE-PHASE POWER

For certain simulations involving three-phase power sources, it may be desirable to connect the power source to a three-phase load at any arbitrary time in the 360 degree input cycle. Load application might occur when phase A is at 0 deg, 90 deg or 270 deg (the three phases will be called A, B and

```
SWITCH.CIR   USE OF VCCS TO MAKE A SWITCH
VPULSE 100 0 PULSE(0 1 1 1M 1M 1)
RPULSE 100 0 1
VL      1 0 10
RL      2 3 20
RMID 3  0 5
RR      3 4 5
VR      4 0 2
GSWITCH 1 2 POLY(2) 100 0 1 2 0 0 0 0 1E6
.TRAN 0.1 2.5
.PLOT TRAN V(3) (0,5)  V(100) (-1,1)
.OPTIONS NOPAGE
.END
```

Figure B.5 Input File SWITCH.CIR.

C).   In this section the generation of three sinusoidal voltages separated by 120 deg of phase and the simultaneous
switching of all three voltages to a load at an arbitrary
point in the cycle will be illustrated.

The obvious way to generate a sinusoid is to use the independent voltage source SIN function, which was introduced in
Chap. 3.5.1.2.   The SIN function allows for any offset voltage, amplitude, frequency, delay time or damping factor to be
specified.   It does not, however, allow for a value of initial phase angle.   Thus, whenever a sinusoidal voltage begins, it starts at 0 V and 0 deg (a zero-crossing with
positive slope).

In order to achieve a phase shift of 120 deg between
phases A, B and C it is necessary to stagger their starting
times by using the delay time parameter for each.   Since a 60
Hz sinusoid has a period of 16.66 ms, a 120 deg phase shift
amounts to a 5.555 ms time delay; 11.11 ms of time delay will
produce a phase shift of 240 deg.   120 V, 60 Hz power would
have a peak value of 120(1.414) or 169.7 Vp.   Three-phase
power could be generated by the following element lines

```
VA  1  0  SIN(0  169.7  60  0)
VB  2  0  SIN(0  169.7  60  5.555M)
VC  3  0  SIN(0  169.7  60  11.111M)
```

```
********11/26/87******** Demo PSpice (May 1986) *******23:05:09********


SWITCH.CIR   USE OF VCCS TO MAKE A SWITCH


****     CIRCUIT DESCRIPTION


***************************************************************************

VPULSE  100  0  PULSE(0  1  1  1M  1M  1)
RPULSE  100  0  1
VL      1  0  10
RL      2  3  20
RMID 3  0  5
RR      3  4  5
VR      4  0  2
GSWITCH  1  2  POLY(2)  100  0  1 2  0 0 0 0  1E6
.TRAN  0.1  2.5
.PLOT TRAN V(3)  (0,5)  V(100)  (-1,1)
.OPTIONS NOPAGE
.END
```

Figure B.6   Output File SWITCH.OUT.

```
****     INITIAL TRANSIENT SOLUTION      TEMPERATURE =   27.000 DEG C

   NODE   VOLTAGE     NODE   VOLTAGE      NODE   VOLTAGE     NODE   VOLTAGE

   ( 1)   10.0000    ( 2)    1.0000     ( 3)    1.0000     ( 4)    2.0000

   (100)    .0000

****     TRANSIENT ANALYSIS              TEMPERATURE -   27.000 DEG C

LEGEND:

*: V(3)
+: V(100)

     TIME      V(3)

(*)-------------    .000D+00    1.250D+00    2.500D+00    3.750D+00    5.000D+00
                     - - - - - - - - - - - - - - - - - - - - - - - - - -

(+)------------- -1.000D+00   -5.000D-01     .000D+00    5.000D-01    1.000D+00
                     - - - - - - - - - - - - - - - - - - - - - - - - - -

   .000D+00  1.000D+00 .        *   .            +            .            .
  1.000D-01  1.000D+00 .        *   .            +            .            .
  2.000D-01  1.000D+00 .        *   .            +            .            .
  3.000D-01  1.000D+00 .        *   .            +            .            .
  4.000D-01  1.000D+00 .        X   .            +            .            .
  5.000D-01  1.000D+00 .        *   .            +            .            .
  6.000D-01  1.000D+00 .        *   .            +            .            .
  7.000D-01  1.000D+00 .        *   .            +            .            .
  8.000D-01  1.000D+00 .        *   .            +            .            .
  9.000D-01  1.000D+00 .        *   .            +            .            .
  1.000D+00  1.000D+00 .        *   .            +            .            .
  1.100D+00  2.000D+00 .            .    *   .                .            +
  1.200D+00  2.000D+00 .            .    *   .                .            +
  1.300D+00  2.000D+00 .            .    *   .                .            +
  1.400D+00  2.000D+00 .            .    *   .                .            +
  1.500D+00  2.000D+00 .            .    *   .                .            +
  1.600D+00  2.000D+00 .            .    *   .                .            +
  1.700D+00  2.000D+00 .            .    *   .                .            +
  1.800D+00  2.000D+00 .            .    *   .                .            +
  1.900D+00  2.000D+00 .            .    *   .                .            +
  2.000D+00  2.000D+00 .            .    *   .                .            +
  2.100D+00  1.000D+00 .        *   .            +            .            .
  2.200D+00  1.000D+00 .        *   .            +            .            .
  2.300D+00  1.000D+00 .        *   .            +            .            .
  2.400D+00  1.000D+00 .        *   .            +            .            .
  2.500D+00  1.000D+00 .        *   .            +            .            .
                     - - - - - - - - - - - - - - - - - - - - - - - - - -

          JOB CONCLUDED

          TOTAL JOB TIME          14.20
```

Figure B.6 (Continued).

However, this is not three-phase power until 11.111 ms, since each phase starts at a different time. The way to overcome this problem is to switch all three phases on at any time after 11.111 ms. Figure B.7 is a schematic of the circuit that will produce switched three-phase power. Notice the independent voltage source VSW with its grounding resistor RSW connected to node 100; it will determine the time at which all three VCCS switches (GA, GB and GC) close.

The input file is shown in Fig. B.8. The VSW element line describes a pulse that goes from 0 V to 1 V at 16 ms. Each of the three switches, GA, GB and GC, is controlled by the voltage at node 100 as well as the voltage across itself. The "on" resistance is the inverse of the transconductance in the VCVS element line, or 1E-6 ohm.

Although the transient analysis will always start at time zero, the graph of load voltages would be zero until 16 ms. For this reason the optional TSTART parameter of 14.16666 ms was included in the .TRAN control line. Results of the transient analysis will be printed and/or plotted beginning at time TSTART.

The graphs in the output file, Fig. B.9, show the phase-to-ground and phase-to-phase voltages resulting from the tran-

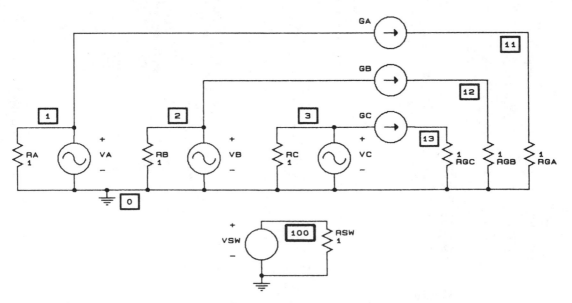

Figure B.7 Three-Phase Power Generator.

```
3-PHASE.CIR    GENERATING SWITCHED 3-PHASE POWER
*       VCCS ARE USED TO SWITCH ALL THREE PHASES ON AT THE SAME TIME
*
*       THE THREE FREE-RUNNING PHASES ARE CREATED BELOW
*
VA      1   0   SIN(0 169.7 60 0)
VB      2   0   SIN(0 169.7 60 5.55555M)
VC      3   0   SIN(0 169.7 60 11.11111M)
RA      1   0   1
RB      2   0   1
RC      3   0   1
*
*       THE SWITCH CONTROL VOLTAGE FOR ALL 3 PHASES FOLLOWS
*
VSW     100 0   PULSE(0 1 16M 1N)
RSW     100 0   1
*
*       THREE VCCS SWITCHES FOLLOW
*
GA      1   11  POLY(2) 100 0  1  11  0 0 0 0 1E6
GB      2   12  POLY(2) 100 0  2  12  0 0 0 0 1E6
GC      3   13  POLY(2) 100 0  3  13  0 0 0 0 1E6
RGA     11  0   1
RGB     12  0   1
RGC     13  0   1
.TRAN   0.83333M   50M   14.16666M
.PLOT   TRAN  V(11) V(12) V(13)
.PLOT   TRAN  V(11,12) V(12,13) V(13,11)
.OPTIONS NOPAGE
.END
```

Figure B.8 Input File 3-PHASE.CIR.

sient analysis. The first graph shows phase A (asterisks) at zero deg, phase B (plus signs) at -120 deg, and phase C (equal signs) at -240 (or +120) deg. The second graph (the phase-to-phase voltages) reveals that there is still a 120 deg shift between V(AB), V(BC) and V(CA). The peak voltage is also higher in the phase-to-phase voltage graph, as would be expected because

$$V(phase\text{-}phase) = V(phase\text{-}ground) * (3)^{0.5}$$

In this example, three voltages were switched just before phase A reached 0 deg. If the switching were to be when phase A was at 90 deg, the pulse VSW would go from 0 V to 1 V at 16.6666 ms plus one-quarter cycle (16.6666 ms/4 = 4.1666 ms), or 20.8333 ms.

```
********11/28/87******** Demo PSpice (May 1986) *******15:40:55********

3-PHASE.CIR   GENERATING SWITCHED 3-PHASE POWER

****    CIRCUIT DESCRIPTION

*****************************************************************************


*     VCCS ARE USED TO SWITCH ALL THREE PHASES ON AT THE SAME TIME
*
*     THE THREE FREE-RUNNING PHASES ARE CREATED BELOW
*
VA      1   0   SIN(0 169.7 60 0)
VB      2   0   SIN(0 169.7 60 5.55555M)
VC      3   0   SIN(0 169.7 60 11.11111M)
RA      1   0   1
RB      2   0   1
RC      3   0   1
*
*     THE SWITCH CONTROL VOLTAGE FOR ALL 3 PHASES FOLLOWS
*
VSW     100  0  PULSE(0  1  16M 1N)
RSW     100  0  1
*
*     THREE VCCS SWITCHES FOLLOW
*
GA      1  11 POLY(2) 100 0  1  11  0 0 0 0 1E6
GB      2  12 POLY(2) 100 0  2  12  0 0 0 0 1E6
GC      3  13 POLY(2) 100 0  3  13  0 0 0 0 1E6
RGA     11   0   1
RGB     12   0   1
RGC     13   0   1
.TRAN   0.83333M   50M  14.16666M
.PLOT   TRAN  V(11) V(12) V(13)
.PLOT   TRAN  V(11,12) V(12,13) V(13,11)
.OPTIONS NOPAGE
.END

****    INITIAL TRANSIENT SOLUTION      TEMPERATURE =  27.000 DEG C

  NODE   VOLTAGE       NODE   VOLTAGE       NODE   VOLTAGE      NODE   VOLTAGE


(  1)    .0000    (  2)     .0000     (  3)     .0000    ( 11)      .0000

( 12)    .0000    ( 13)     .0000     (100)     .0000
```

Figure B.9 Output File 3-PHASE.OUT.

```
****      TRANSIENT ANALYSIS                TEMPERATURE =   27.000 DEG C
LEGEND:
*: V(11)
+: V(12)
=: V(13)

    TIME        V(11)
(*+=)----------  -2.000D+02   -1.000D+02    .000D+00    1.000D+02    2.000D+02
- - - - - - - - - - - - - - - - - - - - - - - - - - - - - - - - - - - -
1.417D-02    .000D+00 .              .          X           .          .
1.500D-02    .000D+00 .              .          X           .          .
1.583D-02    .000D+00 .              .          X           .          .
1.667D-02  -2.831D-04 .        +     .          *      .         =      .
1.750D-02   5.206D+01 .      +       .          .    *      .  =        .
1.833D-02   9.886D+01 .    +         .          .  =    *   .           .
1.917D-02   1.361D+02 .      +       .          . =       .   *         .
2.000D-02   1.602D+02 .          +   .       =  .         .      *      .
2.083D-02   1.690D+02 .        . X   .          .         .       *     .
2.167D-02   1.612D+02 .            = .      +   .         .      *      .
2.250D-02   1.365D+02 .         =    .         .  +       .   *         .
2.333D-02   9.894D+01 .    =         .          .      .+  *           .
2.417D-02   5.196D+01 .     =        .          .      * .  +          .
2.500D-02  -4.131D-02 .       =      .          .  *    .      +       .
2.583D-02  -5.224D+01 .         =  . .      *   .         .        +    .
2.667D-02  -9.970D+01 .            *  .  =       .         .        +   .
2.750D-02  -1.365D+02 .       *      .          =.         .       +    .
2.833D-02  -1.601D+02 .     #        .          .       =  .      +     .
2.717D-02  -1.682D+02 .    *         .          .         .X .           .
3.000D-02  -1.601D+02 .    *         .          .       +  .    =        .
3.083D-02  -1.365D+02 .       *      .          . +       .        =     .
3.167D-02  -9.970D+01 .          *  + .          .         .          =  .
3.250D-02  -5.225D+01 .          + .     *       .         .         =   .
3.333D-02  -5.010D-02 .       +      .          *          .       =     .
3.417D-02   5.195D+01 .    +         .          .      * . = .            .
3.500D-02   9.893D+01 .    +         .          .       = *  .           .
3.583D-02   1.365D+02 .      +       .          . =       .   *          .
3.667D-02   1.612D+02 .          +   .       =  .         .      *       .
3.750D-02   1.690D+02 .         . X  .          .         .       *      .
3.833D-02   1.602D+02 .         =    .          . +       .      *       .
3.917D-02   1.361D+02 .      =       .          .  +      .   *          .
4.000D-02   9.887D+01 .    =         .          .      +  *  .           .
4.083D-02   5.207D+01 .    =         .          .      *   .  +          .
4.167D-02  -1.971D-02 .       =      .          *          .      +      .
4.250D-02  -5.222D+01 .        =   .       *    .         .          +   .
4.333D-02  -9.899D+01 .         *    .          .         .          +   .
4.417D-02  -1.360D+02 .      *       .        =  .         .        +     .
4.500D-02  -1.600D+02 .   *          .          . =       .      +       .
4.583D-02  -1.686D+02 .  *           .          .         .X .            .
4.667D-02  -1.609D+02 .  *           .          . +       .    =         .
4.750D-02  -1.369D+02 .      *       .          +.         .       =     .
4.833D-02  -9.913D+01 .          *  + .          .         .        =    .
4.917D-02  -5.201D+01 .          + . .     *    .         .        =     .
5.000D-02  -9.496D-03 .        +     .          *          .      =      .
- - - - - - - - - - - - - - - - - - - - - - - - - - - - - - - - - - - -
```

Figure B.9 (Continued).

```
****    TRANSIENT ANALYSIS              TEMPERATURE =   27.000 DEG C
LEGEND:
*: V(11,12)
+: V(12,13)
=: V(13,11)
    TIME      V(11,12)
(*+=)----------- -4.000D+02   -2.000D+02    .000D+00    2.000D+02    4.000D+02
- - - - - - - - - - - - - - - - - - - - - - - - - - - - - - - - - - - - - - - -
  1.417D-02   .000D+00 .            .           X          .          .
  1.500D-02   .000D+00 .            .           X          .          .
  1.583D-02   .000D+00 .            .           X          .          .
  1.667D-02  1.469D+02 .     +      .           .        X .          .
  1.750D-02  2.170D+02 .       +    .           .      =  .*         .
  1.833D-02  2.662D+02 .         +  .       =   .          .   *     .
  1.917D-02  2.897D+02 .          . +  =        .          .     *   .
  2.000D-02  2.853D+02 .          =     +       .          .     *   .
  2.083D-02  2.535D+02 .       =     .          .  +       .  *      .
  2.167D-02  1.965D+02 .    =        .          .      +   .*        .
  2.250D-02  1.190D+02 .    =        .          .     * +  .         .
  2.333D-02  3.053D+01 .      =      .          .   *      . +       .
  2.417D-02 -6.056D+01 .        =.   .       *  .          .  +      .
  2.500D-02 -1.459D+02 .         .  X .          .         .  +      .
  2.583D-02 -2.175D+02 .       *.    .       =   .          . +      .
  2.667D-02 -2.684D+02 .      *      .          .    =      . +      .
  2.750D-02 -2.908D+02 .    *        .          .        =  + .      .
  2.833D-02 -2.852D+02 .    *        .          .         +  .  =    .
  2.917D-02 -2.522D+02 .      *      .          .  +       .     =   .
  3.000D-02 -1.950D+02 .        *    . +        .          .       = .
  3.083D-02 -1.188D+02 .          . + *         .          .       = .
  3.167D-02 -3.071D+01 .        +   .          *.          .     =   .
  3.250D-02  6.073D+01 .     +      .          .        *  .    =    .
  3.333D-02  1.457D+02 .     +      .          .        X  .         .
  3.417D-02  2.164D+02 .     +      .          .      =   .*         .
  3.500D-02  2.663D+02 .       +  . .          .   =      .  *       .
  3.583D-02  2.906D+02 .          . +  =       .          .     *   .
  3.667D-02  2.871D+02 .          =     +      .          .     *   .
  3.750D-02  2.536D+02 .       =    .          .  +       .   *     .
  3.833D-02  1.953D+02 .    =       .          .       +  . *       .
  3.917D-02  1.185D+02 .    =       .          .     *  + .         .
  4.000D-02  3.042D+01 .     =      .          .  *       .  +      .
  4.083D-02 -6.081D+01 .       =.   .       *  .          .   +     .
  4.167D-02 -1.467D+02 .         .  X .          .         .   +    .
  4.250D-02 -2.176D+02 .       *.    .       =   .          .   +    .
  4.333D-02 -2.666D+02 .      *      .          .    =      .    +   .
  4.417D-02 -2.897D+02 .    *        .          .       =  + .      .
  4.500D-02 -2.850D+02 .    *        .          .         +  .  =    .
  4.583D-02 -2.528D+02 .      *      .          .  +       .    =    .
  4.667D-02 -1.961D+02 .        *    . +        .          .      =  .
  4.750D-02 -1.193D+02 .          . + *         .          .      =  .
  4.833D-02 -3.062D+01 .        +   .          *.          .    =    .
  4.917D-02  6.052D+01 .     +      .          .        *  .  .=     .
  5.000D-02  1.469D+02 .     +      .          .        X  .         .
- - - - - - - - - - - - - - - - - - - - - - - - - - - - - - - - - - - - - - - -
          JOB CONCLUDED
          TOTAL JOB TIME           30.10
```

Figure B.9 (Continued).

Appendix C

NONLINEAR (POLYNOMIAL) DEPENDENT SOURCES

C.1 NONLINEAR DEPENDENT SOURCES

In Chap. 5 the topic of linear dependent sources was covered. SPICE also supports nonlinear dependent sources, which have great utility for modeling nonlinear resistors, multipliers (including amplitude modulation generators) and voltage-controlled and current-controlled switches, to name a few. The usefulness of nonlinear dependent sources is limited by one's imagination (and patience with mastering the topic).

SPICE recognizes the same four types of dependent sources as mentioned in Chap. 5:

voltage-controlled current source, or VCCS
voltage-controlled voltage source, or VCVS
current-controlled current source, or CCCS
current-controlled voltage source, or CCVS.

The sources are uniquely specified by an element name beginning with the letter G, E, F or H and have units as described below:

ELEMENT	EQUATION	ELEMENT TYPE	UNIT
VCCS	I = G (V)	Transconductance	Siemen
VCVS	V = E (V)	Voltage gain	V/V
CCCS	I = F (I)	Current gain	A/A
CCVS	V = H (I)	Transresistance	Ohm

A nonlinear dependent source can, depending on how it is specified, be very linear. Perhaps a better name would be a polynomial source, since the equations that determine the source output are described by one or more polynomial coefficients. The set of coefficients is p0, p1, . . . , pn. The meaning of a coefficient depends on how many controlling sources there are upon which the polynomial source depends. There is no limit on the number of controlling sources except one's own ability to keep track of the coefficients. My limit is three, except on Mondays it is two.

C.1.1 One–Dimensional Polynomial Equation

If a polynomial source is itself a function of only one controlling source, it is said to be one-dimensional. An expression for the polynomial source, called fv (for function value) in terms of the controlling source, called fa, is

$$fv = p0 + p1*fa + p2*fa^2 + p3*fa^3 + p4*fa^4 + \ . \ . \ .$$

If this were a VCVS, then the unit of p0 would be volt, the unit of p1 would be volt/volt, the unit of p2 would be volt/(volt*volt), etc. In order to make it easy to write element lines for one-dimensional polynomial sources, if one and only one coefficient is specified SPICE will interpret it to be the p1 coefficient, and assumes p0 = 0.0.

C.1.2 Two–Dimensional Polynomial Equation

If a polynomial source is itself a function of two controlling sources, it is said to be two-dimensional. An expression for the polynomial source, called fv (for function value) in terms of the controlling sources, called fa and fb, is

$$fv = p0 + p1*fa + p2*fb + p3*fa^2 + p4*fa*fb + p5*fb^2$$
$$p6*fa^3 + p7*fa^2*fb + p8*fa*fb^2 + p9fb^3 + \ldots$$

If this were a VCCS, then the unit of p0 would be amp, the unit of p1 and p2 would be amp/volt, the unit of p3, p4 and p5 would be amp/(volt*volt), and the unit of p6, p7, p8 and p9 would be amp/(volt*volt*volt).

C.1.3 Three–Dimensional Polynomial Equation

If a polynomial source is itself a function of three controlling sources, it is said to be three-dimensional. An expression for the polynomial source, called fv (for function value) in terms of the controlling sources, called fa, fb and fc is

$$fv = p0 + p1*fa + p2*fb + p3*fc + p4*fa^2 + p5*fa*fb +$$
$$p6*fa*fc + p7*fb^2 + p8*fb*fc + p9*fc^2 + p10*fa^3 +$$
$$p11*fa^2*fb + p12*fa^2*fc + p13*fa*fb^2 + p14*fa*fb*fc$$
$$+ p15*fa*fc^2 + p16*fb^3 + p17*fb^2*fc + p18*fb*fc^2$$
$$+ p19*fc^3 + p20*fa^4 + \ldots$$

If this were a CCCS, then the unit of p0 would be amp, the unit of p1, p2 and p3 would be amp/amp, the unit of p4 through p9 would be amp/(amp*amp), and the unit of p10 through p19 would be amp/(amp*amp*amp).

Once I tried figuring out the coefficients for a four-dimensional polynomial source and my pencil started smoking. If all you are doing is adding four controlling sources, then p0 = 0 and p1, p2, p3 and p4 would be 1. Beyond that, proceed with caution.

C.2 NONLINEAR VOLTAGE-CONTROLLED CURRENT SOURCES

The general form for a nonlinear VCCS is

    GXXXXXX N+ N- <POLY(ND)> NC1+ NC1- <NC2+ NC2-> ...
    + P0 P1 ... <IC = ... >

where N+ and N- are the nodes to which the current source is connected. POLY(ND) must be specified if the VCCS has more than one dimension (the default is POLY(1)). NC+ and NC- are the positive and negative controlling nodes, respectively, and there must be a pair for each dimension. The initial condition, IC = , is the value of the controlling voltages at time zero. If initial conditions are not specified the default value is zero.

EXAMPLE:

GSIMPLE 1 4 7 2 0 0.5 0.3 .06
        The controlled current that flows from node 1 to node 4 through GSIMPLE is a function of the voltage at node 7 compared to node 2, and has the following equation

$$I = 0 + 0.5*V(7,2) + 0.3*V(7,2)^2 + (6E\text{-}2)*V(7,2)^3 \quad amp$$

Since no POLY(ND) is specified, SPICE assumes GSIMPLE to be a function of one voltage only.

EXAMPLE:

GGOLLY 5 9 POLY(2) 1 0 3 4 .001 7M 80U
        Voltage-controlled current source GGOLLY is connected to nodes 5 and 9 (with positive current assumed to flow from 5 to 9 through GGOLLY) and is controlled by two voltages. The first controlling voltage is the node 1 to node 0 voltage, the second is the voltage at node 3 compared to node 4. P0 = 1E-3, P1 = 7E-3 and P2 = 8E-5. This means that the expression for the current is

$$I = 1E\text{-}3 + (7E\text{-}3)*V(1,0) + (8E\text{-}5)*V(3,4) \quad amp$$

EXAMPLE:

It is possible to model a nonlinear resistor by using a nonlinear voltage-controlled current source of one dimension. Consider a VCCS whose controlling nodes are the same as the nodes between which the current flows. The current is then a nonlinear function of the voltage; in other words, a nonlinear resistor. Examine the SPICE input file shown in Fig. C.1. GNLR is a VCCS connected between nodes 8 and 0,

```
NONLINR.CIR  TO MODEL A NONLINEAR RESISTOR
*       A NONLINEAR VCCS IS USED TO MAKE A
*       RESISTOR WHOSE I-V GRAPH CURVES
VSWEEP 5 0 DC
VSENSE 5 8 0
*       IN THE VCCS BELOW (GNLR), THE
*       CURRENT THAT FLOWS BETWEEN NODES
*       8 AND 0 IS CONTROLLED BY THE
*       VOLTAGE BETWEEN NODES 8 AND 0.
GNLR   8 0 8 0 0 1 4
.DC  VSWEEP 0 1 .05
.PLOT DC I(VSENSE) (0,5)
.OPTIONS NOPAGE
.END
```

Figure C.1 Input File NONLINR.CIR.

whose controlling nodes are 8 and 0. The equation for the current through GNLR is

$$I = 1*V + 4*V^2 \quad amp$$

where V is the voltage V(8,0), and I is the current flowing from node 8 to node 0 through GNLR. Figure C.2 illustrates the VCCS and a test circuit to generate an I-V graph.

The graph in the output file of Fig. C.3 shows that as the DC voltage source VSWEEP increases from 0 to 5 V, the current I(VSENSE) increases from 0 to 5 A. However, the slope is not constant; GNLR behaves as a voltage-dependent, nonlinear resistor.

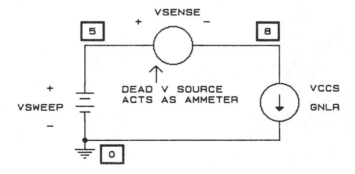

Figure C.2 VCCS Used to Make Nonlinear Resistor.

```
********12/ 2/87******** Demo PSpice (May 1986) *******16:51:38********


NONLINR.CIR   TO MODEL A NONLINEAR RESISTOR


****     CIRCUIT DESCRIPTION


*************************************************************************

*     A NONLINEAR VCCS IS USED TO MAKE A
*     RESISTOR WHOSE I-V GRAPH CURVES
VSWEEP  5  0  DC
VSENSE  5  8  0
*     IN THE VCCS BELOW (GNLR), THE
*     CURRENT THAT FLOWS BETWEEN NODES
*     8 AND 0 IS CONTROLLED BY THE
*     VOLTAGE BETWEEN NODES 8 AND 0.
GNLR    8  0  8  0  0  1  4
.DC  VSWEEP  0  1  .05
.PLOT DC I(VSENSE) (0,5)
.OPTIONS NOPAGE
.END

****     DC TRANSFER CURVES              TEMPERATURE =   27.000 DEG C

    VSWEEP      I(VSENSE)

                   .000D+00     1.250D+00     2.500D+00     3.750D+00     5.000D+00

     .000D+00    .000D+00 *          .             .             .             .
    5.000D-02  6.000D-02 .*          .             .             .             .
    1.000D-01  1.400D-01 .*          .             .             .             .
    1.500D-01  2.400D-01 . *         .             .             .             .
    2.000D-01  3.600D-01 .   *       .             .             .             .
    2.500D-01  5.000D-01 .    *      .             .             .             .
    3.000D-01  6.600D-01 .     *     .             .             .             .
    3.500D-01  8.400D-01 .      *    .             .             .             .
    4.000D-01  1.040D+00 .       * . .             .             .             .
    4.500D-01  1.260D+00 .        *  .             .             .             .
    5.000D-01  1.500D+00 .         * .             .             .             .
    5.500D-01  1.760D+00 .          .  *           .             .             .
    6.000D-01  2.040D+00 .          .   *  .       .             .             .
    6.500D-01  2.340D+00 .          .      * .     .             .             .
    7.000D-01  2.660D+00 .          .       .  *   .             .             .
    7.500D-01  3.000D+00 .          .       .    * .             .             .
    8.000D-01  3.360D+00 .          .       .      .  *          .             .
    8.500D-01  3.740D+00 .          .       .      .    *        .             .
    9.000D-01  4.140D+00 .          .       .      .      .  *   .             .
    9.500D-01  4.560D+00 .          .       .      .      .    * .             .
    1.000D+00  5.000D+00 .          .       .      .      .      .             *

            JOB CONCLUDED

            TOTAL JOB TIME              5.50
```

Figure C.3 Output File NONLINR.OUT.

C.3 NONLINEAR VOLTAGE-CONTROLLED VOLTAGE SOURCES

The general form for a nonlinear VCVS is

EXXXXXX N+ N- <POLY(ND)> NC1+ NC1- <NC2+ NC2-> ...
+ P0 P1 ... <IC = . . .>

where N+ and N- are the nodes to which the voltage source is connected. POLY(ND) must be specified if the VCVS has more than one dimension (the default is POLY(1)). NC+ and NC- are the positive and negative controlling nodes, respectively, and there must be a pair for each dimension. The initial condition, IC = , is the value of the controlling voltages at time zero. If initial conditions are not specified the default value is zero.

EXAMPLE:

ESIMPLE 1 4 7 2 0 0.5 0.3 .06
        The voltage across nodes 1 and 4 (produced by VCVS ESIMPLE) is a function of the voltage at node 7 compared to node 2, and has the following equation

$$V(1,4) = 0.5*V(7,2) + 0.3*V(7,2)^2 + (6E-2)*V(7,2)^3 \quad \text{volt}$$

Since no POLY(ND) is specified, SPICE assumes ESIMPLE to be a function of one voltage only.

EXAMPLE:

EPRODUCT 5 7 POLY(2) 20 22 31 33 0 0 0 0 1
        VCVS EPRODUCT produces a voltage between nodes 5 and 7 (positive and negative, respectively) which is the product of the voltages across node pairs 20 & 22 and 31 & 33. Thus, the equation for EPRODUCT is

$$V(5,7) = 1.0*V(20,22)*V(31,33) \quad \text{volt}$$

This can be used to make amplitude modulation, either double sideband full carrier (DSB-FC) or double sideband suppressed carrier (DSB-SC). To do this, one of the controlling voltages would be a sinusoid at the carrier frequency, and the other controlling voltage would be the modulating signal plus a constant (for DSC-FC) or without the constant (for DSB-SC).

C.4 NONLINEAR CURRENT–CONTROLLED CURRENT SOURCES

The general form for a nonlinear CCCS is

        FXXXXXX N+ N- <POLY(ND)> VN1 <VN2 . . . >
      + P0 P1 . . . <IC = . . . >

where N+ and N- are the nodes to which the current source is connected.   Positive current is understood to flow from N+ to N- through the CCCS.   POLY(ND) must be specified if the CCCS has more than one dimension (the default is POLY(1)).   VN1 VN2 . . .   are the names of the voltage sources through which the controlling currents flow, and there must be one for each dimension.   The initial condition, IC = , is the value of the controlling currents at time zero.   If initial conditions are not specified the default value is zero.

EXAMPLE:

        FUNKY 9 3 VINPUT 1 0.4 2
            Current-controlled current source FUNKY is connected between nodes 9 and 3, and is controlled by the current through independent voltage source VINPUT.   The equation for the current through FUNKY is

$$I = 1.0 + 0.4*I(VINPUT) + 2*(I(VINPUT))^2   amp$$

Since no POLY(ND) is specified, SPICE assumes FUNKY to be a function of one voltage only.

EXAMPLE:

        FUZZY 6 8 POLY(3) VA VB VC 0 2 3 4
            CCCS FUZZY causes a current to flow from node 6 to node 8, through FUZZY.   The equation for the current through FUZZY is

$$I = 2*VA + 3*VB + 4*VC   amp$$

If the P1, P2 and P3 coefficients (2, 3 and 4, respectively) were all made 1, then the current through FUZZY would be the sum of the currents through VA, VB and VC.

C.5 NONLINEAR CURRENT-CONTROLLED VOLTAGE SOURCES

The general form for a nonlinear CCVS is

```
HXXXXXX  N+ N- <POLY(ND)> VN1 <VN2 ... >
+ P0  P1  ...  <IC = ... >
```

where N+ and N- are the nodes to which the current source is connected. Positive current is understood to flow from N+ to N- through the CCVS. POLY(ND) must be specified if the CCVS has more than one dimension (the default is POLY(1)). VN1 VN2 . . . are the names of the voltage sources through which the controlling currents flow, and there must be one for each dimension. The initial condition, IC = , is the value of the controlling currents at time zero. If initial conditions are not specified the default value is zero.

EXAMPLE:

```
HITHERE  7  1  VX  0  2  0  4E2
```
Current-controlled voltage source HITHERE is connected between nodes 7 and 1, and is controlled by the current through independent voltage source VX. The equation for the voltage across HITHERE is

$$V(7,1) = 0 + 2*I(VX) + 4E2*(I(VX))^3 \quad \text{volt}$$

Since no POLY(ND) is specified, SPICE assumes HITHERE to be a function of one voltage only.

Appendix D

BIBLIOGRAPHY

GENERAL:

Barros, Greg and Joe Domitrowich, "Technology Update: Varied circuit simulators led by Spice," *EDN Product News*, October 1986, pp. 12, 31.

Blume, Wolfram, "Computer Circuit Simulation," *Byte*, July 1986, pp. 165-70.

Chua, Leon O. and Pen-Min Lin, *Computer-aided Analysis of Electronic Circuits: Algorithms & Computational Techniques.* Englewood Cliffs, N.J.: Prentice-Hall, Inc., 1975.

Cuthbert Jr., Thomas R., *Circuit Design Using Personal Computers.* New York, N.Y.: John Wiley & Sons, Inc., 1983.

Epler, Bert, "Circuit Simulation and Modeling Column: SPICE2 Application Notes for Dependent Sources," *IEEE Circuits and Devices Magazine*, Vol. 3, no. 5 (September 1987), pp. 36-44.

296

Hines, J. Richard, "Selecting a Personal Computer For Circuit Simulation," *VLSI Systems Design*, Vol. VIII, no. 3 (March 1987), pp. 66-71.

Hines, J. Richard, "Reduce simulation times, costs with SPICE convergence aids," *Personal Engineering & Instrumentation News*, May 1987, pp. 47-51.

Nagel, Lawrence W., "SPICE2: A Computer Program to Simulate Semiconductor Circuits," *ERL Memo No. ERL-M520*, Electronics Research Laboratory, University of California, Berkeley, May 1975.

Rao, Veerendra and Ronald Hoelzeman, "A SPICE Interactive Graphics Preprocessor," *IEEE Transactions on Education*, Vol. E-29, no. 3 (August 1986), pp. 150-53.

Schreier, Paul G., "Simulators benefit from graphic interfaces, reliable convergence," *Personal Engineering & Instrumentation News*, January 1987, pp. 35-43.

Shear, David, "Board-level analog CAE," *EDN*, May 14, 1987, pp. 138-50.

Vlach, Jiri and Kishore Singhal, *Computer Methods for Circuit Analysis and Design*. New York, N.Y.: Van Nostrand Reinhold Company, Inc., 1983.

Yang, Ping, "Circuit Simulation and Modeling," *IEEE Circuits and Devices Magazine*, Vol. 3, no. 4 (July 1987), pp. 40-41.

SEMICONDUCTOR MODELING:

Antognetti, P. and G. Massobrio, *Semiconductor Device Modeling with SPICE*, New York:McGraw-Hill Book Company, Inc., 1988.

Bowers, James C. and H.A. Neinhaus, "SPICE2 Computer Models for HEXFETs", Application Note 954A, *HEXFET Power MOSFET Databook, HDB-3*, El Segundo, CA.: International Rectifier Corporation pp. A-153-60.

Bowers, James C., "Computer design for power electronics," *IEEE Potentials*, May 1986, pp. 36-39.

Early, J.M., "Effects of Space-Charge Layer Widening in Junction Transistors," *Proceedings IRE*, Vol. 46, November 1952, pp. 1401-06.

Ebers, J.J. and J.L. Moll, "Large Signal Behavior of Junction Transistors," *Proceedings IRE*, Vol. 42, December 1954, pp. 1761-72.

Fay, Gary and Judy Sutor, "A Power FET SPICE Model From Data-Sheet Specs," *Powertechnics Magazine*, August 1986, pp. 25-31.

Fay, Gary and Judy Sutor, "Power FET SPICE models are easy and accurate," *Powertechnics Magazine*, Vol. 3, no. 8 (August 1987), 16-21.

Gummel, H.K. and H.C. Poon, "An Integral Charge Control Model for Bipolar Transistors," *Bell System Technical Journal*, Vol. 49, May/June 1970, pp. 827-852.

Herskowitz, Gerald J. and Ronald B. Schilling, *Semiconductor Device Modeling For Computer-Aided Design*, New York: McGraw-Hill Book Company, Inc., 1972.

Macnee, Alan B., "Computer-Aided Optimization of Transistor Model Parameters," *IEEE Transactions on Education*, Vol. E-28, no. 1 (February 1985), 4-11.

Schichman, H. and D.A. Hodges, "Modeling and Simulation of Insulated-Gate Field-Effect Transistor Switching Circuits," *IEEE Journal of Solid-State Circuits*, Vol. SC-3, September 1968, pp. 285-89.

SWITCHING POWER SUPPLIES:

Bello, Vincent G., "Computer program adds SPICE to switching-regulator analysis," *Electronic Design*, March 5, 1981, pp. 89-95.

Appendix E

SOURCES OF SPICE-BASED CIRCUIT ANALYSIS SOFTWARE

1. SPICE, Version 2G.6 and Version 3.

> Ms. Cindy Manly
> EECS/ERL Industrial Support Office
> 497 Cory Hall
> University of California
> Berkeley, CA   94720   (415) 643-6687

The development and writing of SPICE was done at the University of California, Berkeley in the early 1970s. SPICE2G.6 is written in FORTRAN, while SPICE3 is based directly on SPICE2G.6 and is written in the C programming language.   Though they do not support SPICE, they can provide a substantial number of publications on the subject of SPICE and related topics.   Information on availability can be obtained from the address listed above.   A partial listing of the offerings available, and the prices (as of the spring of 1987) are:

SPICE2G.6 User's Guide	$5.00
SPICE3A.7 User's Guide	$5.00
SPICE3A.7 User's Manual	$5.00

SPICE3A.7 Programmer's Manual $5.00
M520: L. Nagel, SPICE2:A Computer
    Program to Simulate Semiconductor
    Circuits, May, 1975. $30.00
  M592: E. Cohen, Program Reference
    for SPICE2, June, 1976. $15.00

SPICE2G.6 source code is available from certain re-distributors designated by U.C. Berkeley for the following machine/operating system combinations: CDC, DEC VAX running UNIX, DEC VAX running VMS, Honeywell, IBM (not for PCs), Unisys Corporation, and Prime. For information on where to obtain SPICE2G.6 for a mainframe computer listed, write to the address listed above. SPICE3A.7 is available for VAX/UNIX or VAX/VMS systems directly from U.C. Berkeley.

2. PSpice.

Microsim Corporation
23175 La Cadena Drive
Laguna Hills, CA 92653   (714) 770-3022

PSpice is a version of Berkeley SPICE for the IBM-type personal computer (although Microsim makes software for other computers) which has some differences from Berkeley SPICE2G.6 . Some of the differences are:

Improved convergence algorithms, leading to faster transient analysis results.

No need for dummy voltage sources to be used as "ammeters" to measure current in a branch; PSpice can give the current through a resistor, capacitor, or inductor directly.

PSpice lacks distortion analysis (.DISTO) and certain distortion output variables are not available.

PSpice always uses the trapezoidal integration method; Berkeley SPICE allows the Gear method to be substituted for the default trapezoidal method of integration.

PSpice has a graphics post-processor called "Probe," which amounts to a software oscilloscope. With Probe, one can look at results of a simulation using graphics on the computer monitor, or can obtain high-quality hardcopy on a printer or plotter.

A math co-processor chip is required to use PSpice.

NOTE: An educational version of PSpice (with Probe) is available to educators at no cost from Microsim Corporation. This version has certain limits on circuit size but is a very powerful tool nonetheless. A math co-processor chip is needed only for using Probe with this version.

3. ALLSPICE.

Contour Design Systems, Inc. (formerly ACOTECH)
713 Santa Cruz Avenue, Suite 2
Menlo Park, CA 94025   (415) 325-7999

ALLSPICE is a derivative of U.C. Berkeley SPICE 2G.6 which is designed for an IBM-type PC. It is menu-driven and interactive but can be run in batch mode. It offers a graphics post-processor (called GINGER) and comes in versions for 512K and 640K of PC memory. A demo version is available. A math co-processor chip is required for both the demo and regular versions.

Also available is COMPLIB, an extensive library of models for semiconductors and ICs. COMPLIB versions for the PC and VAX systems can be supplied.

4. IS_SPICE.

Intusoft
P.O. Box 6607
San Pedro, CA 90734-6607   (213) 833-0710

IS_SPICE is for IBM-type personal computers, and offers DC, AC, and transient analysis capability. Intusoft also offers PRE_SPICE and Intu-Scope add-ons, which are useful additions. Intu-Scope is a graphics post-processor utility. One requirement with IS_SPICE is that .PRINT statements must precede .PLOT statements.

5. Z/SPICE.

ZTEC
P.O. Box 737
College Place, WA 99324   (509) 529-7025

Z/SPICE is for IBM-type personal computers, and offers
DC, AC, and transient analysis capability.   It is a
direct translation of Berkeley SPICE 2G.5.

6. SPICE/SPlot Graphics Processor.

California Scientific Software
1159 N. Catalina Ave.
Pasadena, CA 91104   (818) 798-1201

This PC version of Berkeley SPICE 3a7 and 3b1 includes N
and P MESFET's, distributed RC lines and switches.   In
addition to the 2G.6 analysis types, pole-zero analysis
is included.   SPlot is a graphics post-processor which
makes graphs and can drive printers and plotters.

INDEX

A

Absolute value circuit, 218-21
ABSTOL option, 269
AC analysis, see Analysis: AC
ACCT option, 268, 271
Active filter, 6, 111, 183-88
Ammeter, see Dead voltage source
Amplitude modulation, 219-25, 287, 293-94
Analysis:
    AC (.AC), 21, 41-44, 60, 62-64, 73, 152, 166-93, 243, 253, 259
    DC (.DC), 36-40, 61-62, 152, 160-66, 223, 226, 259
    distortion (.DISTO), 61, 67-68, 153, 243-47, 259
    Fourier (.FOUR), 61, 69-70, 75-76, 153, 235-38, 259
    nodal, 3
    noise (.NOISE), 61, 68-69, 153, 253-59
    operating point (.OP), 61, 65-66, 153-57, 238-43, 259
    sensitivity (.SENS), 61, 67, 153, 248-52, 259
    temperature (.TEMP), 61, 70-71, 153, 223-34, 259
    transfer function (.TF), 61, 66, 153, 227, 238-43, 259
    transient (.TRAN), 45-48, 61, 64-65, 153, 193-223, 259, 272-73, 282-83
    types, 12, 60-61, 77-78
Area factor, 97, 98, 100-101, 103-104, 142, 162
Astable, 148, 212-18, 271

B

Balun, 126-34
Bandwidth, 127, 177, 179
Beta, see BF, BR
BF, BJT forward beta, 1, 96, 102-103, 141, 147-48, 165
Bipolar junction transistor (BJT):
    changing model, 146-49
    element line, 100
    model line, 100-103, 186, 211
    parameters, 102
    using CCCS, 56-58
Bistable, 166, 271
Bode plot, 81, 84, 166, 186, 193
BR, BJT reverse beta, 102-103
Breakdown voltage (BV), 98-99, 142
BV, see breakdown voltage

303

Tear out this card and
fill in all necessary
information. Then
enclose this card with
your check or money
order _only_ in an
envelope and mail to:

Book Distribution Center
PRENTICE HALL
Route 59 at
Brook Hill Drive
West Nyack, New York
10995

COMPUTER-AIDED CIRCUIT ANALYSIS USING SPICE—
Walter Banzhaf

Please send the item(s) checked below. **PAYMENT
ENCLOSED (Check or money order only.) The Publisher will
pay all shipping and handling charges.**

____ PSpice® Student Version Disks (2) IBM® PC compatible.
$7.00 each set. (83463-0)

____ PSpice® Student Version Disk (1) MAC II® compatible.
$6.00 each disk. (83462-2)

____ PSpice® Student Version Disk (1) IBM PS/2 compatible.
$6.00 each disk. (83464-8)

NAME _____

DEPT. _____

SCHOOL _____

CITY _____ STATE _____ ZIP _____

NOTE: PROFESSIONAL/REFERENCE BOOKS ARE TAX
DEDUCTIBLE.
Prices subject to change without notice. Please add sales
tax for your area.

Dept. 1   D-CFJB-BE(6)

Tear out this card and
fill in all necessary
information. Then
enclose this card with
your check or money
order _only_ in an
envelope and mail to:

Book Distribution Center
PRENTICE HALL
Route 59 at
Brook Hill Drive
West Nyack, New York
10995

COMPUTER-AIDED CIRCUIT ANALYSIS USING SPICE—
Walter Banzhaf

Please send the item(s) checked below. **PAYMENT
ENCLOSED (Check or money order only.) The Publisher will
pay all shipping and handling charges.**

____ PSpice® Student Version Disks (2) IBM® PC compatible.
$7.00 each set. (83463-0)

____ PSpice® Student Version Disk (1) MAC II® compatible.
$6.00 each disk. (83462-2)

____ PSpice® Student Version Disk (1) IBM PS/2 compatible.
$6.00 each disk. (83464-8)

NAME _____

DEPT. _____

SCHOOL _____

CITY _____ STATE _____ ZIP _____

NOTE: PROFESSIONAL/REFERENCE BOOKS ARE TAX
DEDUCTIBLE.
Prices subject to change without notice. Please add sales
tax for your area.

Dept. 1   D-CFJB-BE(6)